ENERGY & THE ENVIRONMENT

**The University Press
of New England**
Brandeis University
Clark University
Dartmouth College
University of New Hampshire
University of Rhode Island
University of Vermont

ENERGY & THE ENVIRONMENT: A STRUCTURAL ANALYSIS

Edited by Anne P. Carter

Published for the Brandeis University Press
by the University Press of New England
Hanover, New Hampshire   1976

Preface

The studies in this volume were begun at the Harvard Economic Research Project during its final year, 1971–72. They followed the first crude implementation of the augmented input-output system for the study of environmental problems by Leontief and Ford.* Each addresses a specific aspect of pullution or energy economics in greater depth than was possible in the Leontief and Ford across-the-board study.

Because both pullution abatement and the generation of electricity are capital intensive activities, the expanded input-output system was specified in dynamic form in several of the studies. Istvan elaborates the Leontief dynamic inverse by introducing a lag operator matrix that disaggregated capital coefficients according to the construction lag for putting the capital in place. The use of varying construction lags makes for a much more realistic analysis of investment than would be possible with uniform lags. In electric utilities, construction of new plants sometimes takes seven or more years, and the time-phasing of capital inputs into this sector has an important bearing on the timing of demand for investment goods.

In Berlinsky's study of the steel industry, the input-output model is spliced with a meteorological model of the diffusion of particulates that translates emissions into air quality. Since

*Wassily Leontief and Daniel Ford, "Air Pollution and the Economic Structure: Empirical Results of Input-Output Computations," in A. Brody and A. P. Carter, eds., *Input-Output Techniques* (Amsterdam: North-Holland Publishing Company, 1972), pp. 9–30.

emissions affect consumers through deteriorating air quality, a more meaningful interpretation of "final deliveries of pollution" to consumers becomes possible. These and other specific innovations are discussed in some detail in the individual chapters. Chapter 1 assesses the economy-wide implications of the different kinds of energy and pollution control costs for the economy's growth potential.

As the individual studies progressed, we recognized that we were building a comprehensive data base for the study of pollution, covering emissions and abatement processes for several air and water pollutants and solid waste, and that these data might provide a valuable basis for other studies of environmental problems. However, the publication of the volume was delayed by various disruptions that came with the closing of the Harvard Project, delay at the publisher's, and my own overcommitment in too many areas. By the time the book went to press, the Environmental Protection Agency had already constructed a far more comprehensive and up-to-date set of emissions and abatement cost data, and we decided that publication of our own data base would serve no useful purpose.

Although some of our references are a little out of date, the conclusions are still timely. Good ideas, in any case, have a low rate of obsolescence, and we hope that readers will find some that are fresh and interesting.

The initial work done at Harvard was sponsored by grants from the National Science Foundation (Contracts GS-1456X and GI-30021/RANN), and the work continued at Brandeis was sponsored by the National Science Foundation under Contract GS-36000X1. The manuscript was prepared for publication at Brandeis by Arva Rosenfeld Clark. Jane Haynes and Martha Grice also spent a lot of time and energy chasing down far-flung contributors and their graphs and elusive references.

Waltham, Massachusetts   Anne P. Carter
February 1975

Contents

Tables & Figures

1

Introduction:
Energy, Environment,
and Economic Growth

ANNE P. CARTER

1.0 INTRODUCTION

Recent environmental problems and energy shortages have
prompted questions about the future course of economic growth
in the United States. Spokesmen for industry and government
express concern lest capacity bottlenecks interfere with growth
at the 3.5 percent rate that we have sustained over the past quarter century. Others challenge the desirability of the rapid economic growth that we have long taken for granted.

Economic growth is not a simple policy variable. Its future
course will depend on many factors: the availability of natural
resources, now only vaguely appraised, both at home and abroad;
the expansion and composition of the labor force; public and
private consumption; environmental policy; and specific techniques of production. Furthermore, though technology does not

*The author wishes to thank Orani Dixon, Terry Jenkins, and Brian
Dorsey for assistance at earlier stages, and Carol T. Everett, who was responsible for all of the computations at the final stage of analysis. This chapter
was published as an article in *The Bell Journal of Economics and Management Sciences*, Vol. 5, no. 2 (Autumn 1974), pp. 578–592.

uniquely determine our development, it limits the capacity of
the economy to expand with any given consumption and resources. As resource supplies and environmental standards
change, new technologies are developed in response. In principle,
these new technologies could increase or decrease the economy's
growth potential. Actually, the "first generation" technologies
addressed to today's environmental and resource conditions tend
to decrease it. The purpose of this first chapter is to appraise,
roughly and quantitatively, the implications of some specific
pollution-abatement and energy technologies that tend to reduce
the rate of economic growth.

Most of the data for the analysis comes from other studies in
this volume that articulate individual abatement techniques and
energy processes in a dynamic input-output framework [1].
Three sets of innovations are considered: projected changes in
the technologies of generation, transmission, and distribution of
electric power over the next ten years; coal gasification; and a
broad range of pollution-abatement activities that would satisfy
announced or projected standards for air, water, and solid waste
pollution. While far from exhaustive, this is a fairly broad list of
direct technological responses to environmental and resource
conditions expected to come on line over the next decade. Their
adoption could make important differences in our growth potential and in the relative importance of sectoral contributions to
output and to the capital stock. If these changes are introduced,
how will the economy of 1980–85 differ from today's?

In a closed dynamic input-output system, the maximum rate
of economic growth that can be sustained depends on the input
structures of all sectors, both on capital and on current account,
including the level and composition of final consumption. One
important conclusion is that growth potential is more sensitive
to changes in the propensity to consume than to increasing costs
in the energy sectors. The expansion of the economic system is
also constrained implicitly by the availability of labor and specific natural resources. Until very recently the size of the labor
force seemed to be the only major resource constraint that was

binding on the U.S. economy. Since the work force was increasing at only one percent per year, continual increases in productivity were necessary to keep the effective labor supply sufficient for 3–4 percent growth. Thus labor-saving progress dominated technical change during the postwar period. The great bulk of industrial innovations—automation, computerization, material and design changes—represented direct or indirect economies of labor, with direct labor-saving predominating [2]. Had per capita consumption remained constant, increasing labor productivity might have freed capacity for increased capital accumulation and accelerated economic expansion. However, increasing productivity was offset by rising consumption per worker, and thus the average rate of growth was fairly steady. Commoner [3] and others point out that much of the recent increase in labor productivity and in consumption has been achieved through more intensive exploitation of nonhuman resources that once seemed plentiful, particularly energy and the natural environment. Saving labor at the expense of other natural resources may well have hastened the present "crisis" at a time when environmental and energy constraints have also begun to bind.

Just as labor-saving innovations forestall labor shortages, so our environment- and energy-oriented innovations address potential scarcities of natural gas and oil, clean air and water, etc. Precipitators, filters, and treatment plants are designed to protect the air and water; nuclear power and coal gasification are designed to supplement or supplant scarce supplies of clean fossil fuels; higher-voltage transmission and underground distribution lines should convey electric power to consumers with minimal environmental damage. Although these new technologies, once installed, do not use much direct labor, they generally require more capital and intermediate inputs than the old technologies they supercede or augment. These added requirements impose a drag on the growing system.

In sum, two types of influences will tend to limit growth in the coming decade:

(i)   An increased need for labor to maintain current consumption leaves a smaller portion of productive capacity for capital formation;

(ii) Environmental considerations and scarcities of non-human resources complicate the technological challenge of increasing the productivity of labor. Innovators must now deal with environmental problems as well as labor efficiency. It may well prove feasible to increase the efficiency of natural resources and labor simultaneously, but the problem is more difficult than that posed by labor alone. Unless effective labor and resource supplies increase in step with capital accumulation, plans to expand capacity cannot be realized.

To the extent that pressures to conserve natural resources actually reduce the rate of economic growth there will be less pressure to increase labor productivity. If growth of industrial capacity were very slow, natural increases in the labor force alone might provide sufficient manpower for the growing stock of equipment. However, we shall show that minor changes in consumption could readily offset the increased intermediate and capital requirements of the new technologies so as to maintain or even increase the long-run growth rate. Thus it is hard to predict whether the problems of environment and energy supply will tend to relax or to intensify historic pressures for ever-increasing labor productivity.

2.0 COMPUTATIONS

2.1. Analytical Framework

A closed dynamic input-output model (eq. 1) is the basis for all the computations:

$$(I-A)X - B\dot{X} = 0 \tag{1}$$

where X is a vector of total sectoral outputs and $\dot{X}$ is a
vector of their time derivatives.

A and B are current account and capital coefficient matrices
respectively. The model is closed by the addition of a household
row and column to A and B. The household row is a vector of
coefficients representing income and indirect tax payments; the
household column is a vector of coefficients representing expenditures by households and government plus net exports.

Equation 1 says that outputs are allocated to current account
uses, AX, and to expansion, $B\dot{X}$. Since capital formation is
endogenous, the sum of expenditures by the household industry
is equal in national accounts terminology to gross national product minus gross private capital formation. If all sectors were to
grow at a uniform rate, equation 1 could be rewritten

$$(I-A-\lambda B)X = 0 \tag{2}$$

where λ is the uniform or "turnpike" growth rate.

Tsukui [4] has shown that λ represents the unique uniform
growth rate for the economy consistent with full capacity utilization of all sectors. This turnpike growth rate proves to be the
maximum growth rate consistent with the given flow and capital
coefficient matrices for a broad range of initial and terminal
outputs. Corresponding to the turnpike growth rate, λ, is a set
of output proportions, the positive eigenvector of the matrix
$(I-A-\lambda B)$. Thus λ measures the economy's long-run growth
potential with the specified input structures A and B, while the
turnpike output proportions indicate rough norms for the relative importance of sectors on the uniform growth path. Tsukui
[5] and Brody [6] have estimated turnpike paths for Japan and
the United States. Their findings suggest that both these economies operate reasonably close to the computed paths.

2.2 Computation Procedures

The iterative algorithm of Brody [6] was used to compute uniform growth rates and output proportions. We begin with a set of base year output proportions, X°, estimate a trial value of λ and reestimate equilibrium output proportions, X^1, consistent with the first estimate of λ. In outline, the computation sequence is

1. $$\lambda^k = \frac{(1, 1, \ldots, 1)\,(I\text{-}A)X^{k-1}}{(1, 1, \ldots, 1)BX^{k-1}}$$

(3)

2. $X^k = (A + \lambda^k B)X^{k-1}$

where k is the number of the iteration. All computations converged within 7 iterations.*

2.3 Comparative Dynamics

Structural changes in matrices A and B will of course bring changes in the turnpike growth rate and output proportions. In this study the economic impact of structural changes is evaluated in terms of comparative dynamics. We establish benchmark growth rate and output proportions on the basis of capital and current account matrices representing the structure of production in the early 1970's. Successive changes in the base-year matrices are introduced to simulate the adoption of new technologies of energy production and stricter standards of pollution abatement. The effects of these structural changes, singly and in combination, on λ measure their impacts on the potential growth rate of the economy; changes in computed output proportions represent differential impacts of the specified structural changes on individual sectors.

---

*Computations were performed on the PDP-10 computer at the Feldberg Computer Center, Brandeis University, using the PASTIM matrix manipulation program of Richard Drost.

The effects of changes in intermediate input and capital coefficients differ depending on concomitant changes in the levels and patterns of final consumption. To illustrate the impact of changing consumption, alternate consumption vectors were introduced in combination with the changes in industrial coefficients. The alternative consumption patterns are discussed in 2.4.3.

2.4 Implementation

2.4.1 Base Year Matrices

The 83-order 1970 coefficient matrix of the Interagency Growth Project [7] was chosen as the base year A matrix. This coefficient matrix is the result of a set of 1970 projections made in the middle sixties and is not based on actual 1970 statistics. The capital matrix is an update of the 1958 capital matrix of the Harvard Economic Research Project to 1970–75 technologies by the Battelle Memorial Institute [8]. Updating incorporates engineering information and conjecture. That the A and B matrices are both crude projections should not seriously impair their usefulness in this study. They serve primarily as general points of reference for analyzing the impact of specific further changes. Both A and B are in 1958 prices. Annual replacement requirements were estimated and added to the A matrix. The current account and capital matrices were augmented to include a "household" row consisting of the sum of income payments and indirect taxes per dollar of output for each industry and a "household" column representing all purchases of final outputs per dollar of net national income. The "household" sector includes government as well as private consumption expenditures but excludes gross capital formation. The turnpike growth rate computed for the base A and B matrices was 3.5 percent per annum.

2.4.2 Technological Variants

The base matrices were modified by introducing coefficients
that represent the new technologies. Quantitative estimates of
new industrial structures are drawn from specialized studies by
Istvan, Just, Jenkins, Berlinsky, Kok, and Dorsey [1].

Three major sets of structural changes were considered:

(i) The first consists of changes in technology for electric
power generation, transmission, and distribution. The input
coefficients in the base year's flow coefficient matrix were replaced by coefficients for a projected 1980 input structure of
the electric power sector. These new coefficients were based on
the "high-nuclear" scenario estimated by Istvan [9]. The corresponding column of the base-year capital coefficient matrix
was similarly replaced by Istvan's estimates of 1980 capital
coefficients. Istvan's projections show that increases in the relative importance of nuclear generation, additions of pollution
abatement equipment, extra high voltage transmission and
underground distribution lines will result in greater costs and, in
particular, greater capital intensity in electric power delivery. He
projects an increase in the total capital coefficient for the electric power industry from 3.0 to 4.9 and a 9 percent increase in
real costs on current account (including replacement).

(ii) The second set of modifications convert the increment in
projected over present fossil fuel consumption from electric
utilities to gasified coal. If all additional consumption of fossil
fuel took the form of gasified coal, then by 1985 the input coefficients of coal, petroleum, and natural gas into electric utilities
would be cut to half of their former values. These coefficients
were so cut and a coal gasification industry was then added to
the base matrix. Its row has a single entry representing utilities'
consumption of gasified coal equivalent in btu value to the reductions in conventional fossil fuel consumption. The column represents Just's projected input structure for the Hygas process of
coal gasification [10]. This represents the most speculative of
all the new technologies included in this study. Because there is

no on-line experience with Hygas or any of the other possible methods, the coefficient estimates are more speculative than estimates for, say, nuclear power generation, cooling towers or sanitary landfill of solid wastes. Environmental standards will require some process of coal gasification and/or liquifaction as reliance on coal increases. This process represents a feasible medium-cost variant of a number of coal gasification alternatives.

(iii) Over the past quarter century electric power input coefficients into all consuming sectors have been increasing steadily at an average rate of roughly 3.5 percent per year [2] and Federal Power Commission projections of electricity consumption implicitly assume that these trends will continue over the next decades [11]. Increased electricity consumption is likely to intensify the impact of changing energy technologies. Computations representing the technological changes in (i) and (ii) were repeated, with the base year energy row coefficients increased by 40 percent. The adjusted electric power row represents the economy's power consumption propensities a decade hence if present trends continue.

(iv) To stimulate the impact of economy-wide pollution controls, the basic flow and capital coefficient matrices were augmented to include an abatement cost row and a dummy abatement column for each of six types of pollution: particulate air pollution, industrial water pollution, thermal pollution, municipal sewage disposal, strip mining pollution, and municipal solid waste disposal [12]. The levels of abatement activities are calculated to satisfy the abatement standards of the Clean Air Act of 1971 for particulates and requirements for primary treatment (elimination of 85 percent of BOD and suspended solids) of municipal and industrial waste water.

2.3.3 Consumption Variants

In the base matrix consumption proportions are those of 1970, and final consumption grows at the same rate as all other sectors on the turnpike path. Per-capita consumption in the future will depend on the rate of growth of the population relative to the rate of growth of total final goods. In the absence of increases in the work force the economy can expand only as fast as increasing per-capita productivity permits. Under these conditions, per-capita real income increases at the same rate as the economy's uniform growth rate. Should population increase at the same rate as the economy, the rate of increase in per-capita income would be zero. Over the next decade the rate of growth of the population will be positive but probably lower than the growth rate of the economy and rising per-capita income is generally expected. Consumption patterns can be expected to change as per-capita income rises and as product prices change. Because the household sector is large, the structure of consumption has a major influence on growth potential. The impact of changing industrial technologies was computed with each of four different consumption structures.

(i) The first is that of the base year.

(ii) The second makes a rough allowance for expected changes due to rising per-capita income. Ten years hence, assuming 2.5 percent growth per annum in labor productivity, we can expect real income per capita to be 28 percent higher than it is now.* To estimate the effect of this rise in average per-capita income on consumption patterns, we assume that spending patterns will remain the same as they are now for any given income level. But as average income goes up, the relative importance of high-income spending patterns will increase. To estimate one possible structure of future consumption we assume that all incomes will increase proportionately as GNP rises. Each income-specific spending pattern is then assigned a weight equal to the proportion of income received by those who earned 28 percent *less* income in the base year. The most striking difference between

---

*Since the computations are illustrative, the precise increase in income per capita chosen here is somewhat arbitrary. Precise consistency with the computed growth rate could be achieved by an iterative process.

this projected structure of consumption and that of the base year lies in the personal savings rates. Because people with high incomes save much more than those in lower income groups, the percentage rate of savings under variant ii is triple that in the base year. History shows, however, that personal savings remain a remarkably stable percentage of the national income, rising very slowly with time and/or per-capita income. Consumers may be ready to save more in the future if environmental improvement requires it.

(iii) Here the savings rate is allowed to increase by only *one* percentage point over the base pattern (i), in keeping with historical trends. Proportions of final spending on goods and services are kept the same as in variant ii.

(iv) Over the next ten years rising energy prices are likely to induce substitution in favor of products that are less energy-intensive. This variant superimposes energy-saving changes into the base-year consumption patterns. Direct consumption of energy is cut by 20 percent, while expenditures on other consumer items are increased proportionately so as to maintain the base-year savings rate.

Table 1 lists the various scenarios that were computed.

3.0 RESULTS

3.1 Effects of Selected Changes on λ

Table 2 shows the effects on growth potential of specific technological shifts, separately and in combination, with the fixed base-year consumption structure. Few of these changes affect growth potential by more than a few tenths of a percentage point. The cumulative effect of groups of changes is, however, appreciable. Abating all water, air, and solid waste pollution to meet currently announced goals reduces λ from 3.5 to 3.0 percent. Projected

Table 1. *Scenarios Computed in This Study*

Consumption Structure	(i) 1970		(ii) Income-specific consumption structures reweighted for 28% increase in real income. Savings triple base year	(iii) 1% increase in savings over 1970. Consumption pattern of (ii)	(iv) Energy-saving changes. Base year consumption pattern
Electricity Consumption Coefficients	(a) 1970	(b) 1970 X 1.4			
Technology					
1. base year	X	X	X	X	X
9. all pollution abatement	X		X	X	X
10. 1980 electric	X				
11. 1980 electric with coal gasification	X	X			
12. all abatement, 1980 electric and coal gasification		X	X	X	X

changes in electric power generation and distribution technology are introduced separately and also combined with coal gasification. The combined changes lead to a .5 percentage point cut in λ. With pollution controls and changes in electric power generation combined, long-run growth potential is now reduced to 2.6 percent per year.

Table 2. *Long Term Growth Potential* (λ)
with New Energy and Abatement Technologies

Structural Change	λ percent per annum
1. none (base-year structure)	3.54
Pollution Abatement	
2. particulate air pollution ($\simeq 99$ percent)	3.44
3. industrial BOD and suspended solids (primary treatment)	3.47
4. municipal BOD and suspended solids (primary treatment)	3.45
5. acid mine drainage	3.49
6. thermal pollution (cooling towers)	3.52
7. all water pollution	3.33
8. solid waste disposal	3.33
9. all pollution	3.03
New Energy Technologies	
10. 1980 electric power technology	3.32
11. 10 + coal gasification	3.06
12. energy and pollution control (9 + 11)	2.59

3.2 Effects of Identified Changes Combined with Rising Industrial Use of Electric Power

The impact of electric power technology on the growth rate is magnified greatly when electric power usage expands. Table 3 shows how new technologies affect the growth rate when power consumption coefficients are increased. Istvan's projected changes in electric power technology alone reduce growth potential by two tenths of a percentage point. If electric power consumption coefficients are multiplied by 1.4 in the base matrix, λ is reduced by .6 percentage point. If this increase in electric power consumption is used in conjunction with Istvan's changes, λ is reduced by .8 percentage point. The effects of

Table 3. *Long Term Growth Potential* (λ)
With Varying Energy Technologies and Electricity Use
(percent per annum)

| | Electricity Consumption Coefficients | |
| | 1970 | 1970 × 1.4 |
Energy Technology	(a)	(b)
1. base year	3.5	2.9
10. 1980 electric	3.3	2.5
11. 1980 electric with coal gasification	3.1	2.1

increased power consumption are even more dramatic—a reduction of 1.0 percentage point—when the increase takes place in a technological scenario that includes coal gasification as well as Istvan's changes. The costs associated with projected increases in electric power usage are larger than those incurred from new developments in electric power technology. Hence increasing power consumption cuts the growth rate more than the switch to new techniques. Separately, neither of these two developments has an overwhelming impact on growth potential, but their tendency to compound each other is striking. The combined effect, 1.4 percentage points, is certainly large enough to be taken seriously.

3.3 Technological Changes Combined with Changes in Consumption

"Households" is the largest sector of our model, comprising all public and private final consumption. At present its level of consumption is roughly 5 times the volume of gross private domestic investment. Hence it is no surprise that growth potential is more sensitive to a given percentage change in household spending and saving patterns than to the same percentage change in individual

Table 4. *Long Term Growth Potential* (λ)
with Varying Technologies and Consumption
(percent per annum)

Technology	Consumption Structure			
	(i)	(ii)	(iii)	(iv)
1. base year	3.5	5.7	3.8	3.6
9. all pollution abatement	3.0	5.1	3.3	3.1
12. electric, coal gasification and all pollution abatement	2.6	4.7	2.9	2.7

industrial sectors. For each of the four consumption structures discussed in 2.3.3, Table 4 shows computed values of λ assuming all pollution controls, separately and in combination with new electric generating and coal gasification technology.

Table 4 confirms the sensitivity of the economy's growth to changes in consumption patterns. Reducing the energy content of final consumption (iv) greatly attenuates the impact of new energy technologies. Were household savings at any given real-income level to remain constant, as in variant ii, increasing per-capita income would raise savings to a point where 5 or 6 percent growth could be sustained, even in the face of our major "deteriorations" in technology. Greater affluence will mean spending more on entertainment and services, less on food and rent. The effect of such a shift in spending patterns tends to increase growth potential even when savings are restricted to only small percentage-point increases over the base year (variant iii).

Figure 1 shows the tradeoffs between consumption and growth with and without pollution abatement and new energy technologies.

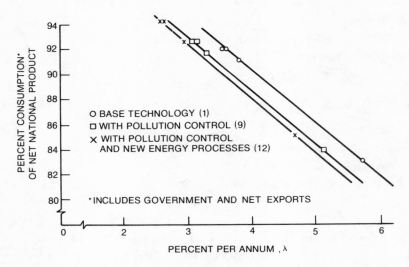

Figure 1
*Tradeoffs between Consumption and Growth
with Changing Technologies*

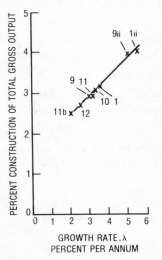

Figure 2. *Importance of
Construction with Changing
Growth Rates*

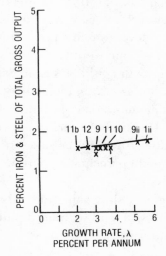

Figure 3. *Importance of Iron
and Steel with Changing
Growth Rates*

3.4 Output Proportions

By changing intermediate demand, new technologies alter the relative importance of individual sectors. Higher capital requirements of new energy and pollution control technologies might be expected to increase the relative importance of capital-producing sectors. On the other hand, deceleration of economic growth means a smaller proportion of economic resources devoted to capital formation. Because these tendencies offset each other, it is difficult to make a priori judgments about the effects of the new technologies on the relative importance of individual sectors. Our computations give some indication of what they may be:

Figures 2 and 3 show the proportionate contributions of the new construction and iron and steel sectors to total gross output under various scenarios that we have computed. Proportion of total gross domestic activity is measured on the vertical axis, while λ is shown on the horizontal axis. The circled number next to each point indicates a combination of technological and consumption structures identified in Tables 1 to 3.

Table 5 shows that the relative importance of new construc-

Table 5. *Sectoral Shares of Total Gross Output with Various Technologies and Consumption Patterns (percent of total gross output)*

	Scenario number*							
	11b	12	9	11	10	1	9(ii)	1(ii)
λ	2.1	2.6	3.0	3.1	3.3	3.5	5.1	5.7
Producing Sector								
New construction	2.5	2.7	2.9	2.9	3.0	3.1	3.9	4.1
Maintenance construction	2.2	2.1	4.4	2.1	2.1	2.2	2.1	2.1
Food	4.4	4.3	4.4	4.4	4.3	4.4	3.8	3.8
Iron and steel	1.6	1.6	1.6	1.6	1.6	1.6	1.8	1.8
Automobiles	3.1	3.1	3.1	3.1	3.1	3.1	3.1	3.1
Electric utilities	2.8	2.2	2.0	2.1	2.2	2.0	1.8	1.8

*Scenarios are listed in Table 1.

tion in the economy varies from 2.5 to 4.1 for different assumptions. Since new construction is the largest single industrial component of most sectors' capital stocks and also of household capital, it is not surprising that this sector's importance varies directly with the growth rate. However, the relative importance of the iron and steel sector is constant because steel is used to make many consumer products (automobiles, cans, household equipment) as well as in capital goods.

The relative importance of most sectors is, like that of steel, virtually invariant with respect to the changes considered in this study. This is demonstrated in Table 5. Utilities increase and decrease in relative importance as electricity input coefficients are varied. Food declines in relative importance as high-income consumption patterns are introduced and the growth rate rises to the neighborhood of 5 percent. By and large, however, the proportions cited for each sector in Table 5 are remarkably stable. This tendency is explained by the limited range of the growth rates covered and the modest proportion of total resources devoted to growth under any of the options considered.

3.5 Changing Composition of Capital Stock

Although the composition of annual output remains relatively stable with respect to our changes, a growing proportion of the capital stock is devoted to energy and abatement capital and the average capital/output ratio for the economy rises by as much as 6 percent over the base year economy. Table 6 shows that by and large higher capital intensity is characteristic of the lower growth-rate scenarios. It also gives the proportion of total capital invested in utilities and in pollution abatement under various sets of assumptions. Utilities accounted for 8.8 percent of total investment in the base year. Initially they have a much greater capital/output ratio than most other sectors. Increasing investment requirements for utilities and increasing use of electric power adds significantly to overall capital requirements.

Table 6. *Economy-Wide Capital/Output Ratios and Proportion of Investment in Utilities and Abatement with Varying Structures (dollars per dollars of total gross output)*

Scenario Number	Growth Rate (λ) (% per annum)	Capital/ Output Ratio	Percent of Total Investment in	
			Utilities	Pollution Abatement
11b	2.1	.84	15.3	*
1b	2.9	.81	12.1	*
9i	3.0	.80	9.8	3.2
11a	3.1	.81	11.8	*
1	3.5	.79	9.9	*
12ii	4.7	.82	10.8	2.5
9ii	5.1	.80	8.9	3.2
1ii	5.7	.78	9.1	*

*Pollution abatement costs are not estimated for these scenarios.
Source: see text.

4.0 CONCLUSIONS

The above results show the sensitivity of the economy—of its sectoral proportions and growth potential—to prospective structural changes. The specific assumptions regarding the rates at which new technologies will be introduced may tend to exaggerate their dampening effects. Gasified coal will probably account for only a small proportion of increases in fossil fuel use over the next ten years. Increasing costs of electricity should slow the rate of increase in electric power consumption. Still, the potential effects are not negligible. Although, taken separately, no one of the technological changes considered in this study exerts a decisive effect on the capital and output proportions of the economy or on its growth potential, their combined effect could be substantial. Furthermore, we have dealt with only a few of the major adjustments that environmental and resource pressures will induce over the next decade. As the level of production gets larger, progressively stricter environmental safeguards will be necessary simply to maintain any given standard of air and water quality. More substances will be recognized as pollutants. Most known abatement technologies involve very sharply increasing costs when the percentage of residual pollution is to be lowered. Significantly higher costs of extraction and refining of all fuels and ores is generally anticipated.

On the other hand, adaptive responses to rising energy and pollution costs soften the impact of these resources and environmental constraints. Our computations show, for example, that a 3.5 percent per year increase in electricity consumption coefficients greatly enhances the growth-limiting effects of new energy technologies. Conversely, a modest reduction in electricity coefficients could neutralize their impact. Ayres and Gutmanis [13] point out that current best-practice technologies in many polluting sectors already show substantially lower gross emissions than average technologies now in use. The cumulative impact of small industrial and consumer adaptations to changing energy and environmental cost conditions is hard to anticipate. It could be very large.

Within the framework of this analysis we could study changing economic options as additional technological changes evolve. Although the relative contributions of individual sectors do not change markedly in the face of the changes just considered, it would be unrealistic to ignore the lags and potential bottlenecks likely to arise in a sequence of major technological transitions. Leontief [14] and Istvan [15] offer a more flexible framework for tracing year-to-year changes in outputs and investment that depart from the turnpike path.

The value of a general equilibrium approach extends beyond the limited area of technological policy to broader economic questions concerning the changing balance of income and tax policy, public spending, prices, and labor productivity. For some years, at least, the economy's growth potential with any given consumption structure will be lower than it used to be. This study has shown, however, that a small once-for-all increase in consumer savings can easily offset the drag, as could other shifts in consumer patterns. Thus there remains a wide range of trade-offs between faster growth and consumption at any given point. Cross-sectional evidence shows that high-income consumers save a substantially greater portion of their incomes than low. Were consumer savings to rise with income over time at even half the rate they do in cross section, the growth rates of the past twenty-five years could easily be sustained. Additional savings might also be achieved through changes in government fiscal policy to produce a budgetary surplus or to favor business saving. Actual future growth will depend on the complex policies that determine real savings for any given technological context.

Fundamentally, the turnpike growth rate measures the rate at which the economy's productive capacity—its capital stock—can be increased. Real growth also requires that the effective labor supply be sufficient to man the growing productive capacity. Increases in labor productivity have contributed more than growth in the labor force to rising output in recent years [16]. Recent studies suggest, however, that the rate of increase of productivity is decelerating [17, 18]. Some of this deceleration may reflect resource pressures. Whatever its explanation, lagging productivity change could also limit growth. Here, as in the case of consumer savings, the economy retains a broad range of options. While both population and productivity increases are decelerating, rising labor force participation can easily offset their effect on growth potential. At a time when women are clamoring for a share in the job market, youth suffer high rates of unemployment, and firms experiment with the four-day week, labor shortages need not obstruct growth—not yet anyway.

Pollution abatement might be viewed as a public good—i.e., as a form of consumption, but our conclusions do not rest on this assumption. We can meet environmental standards and resource constraints over the next decade and still maintain or even increase present growth rates for conventional goods. This should not be construed as advocacy of rapid economic expansion. The social and environmental evils of bigness are not all captured and dealt with in pollution abatement vectors; nor is it possible to represent all the trade-offs between the present and the remote future, at home and abroad, in a growth model. Despite present constraints we can still grow. Whether we should or want to is another question.

REFERENCES

1. See Chapters 2-6.
2. Carter, Anne P., *Structural Change in the American Economy* (Cambridge: Harvard University Press, 1970).
3. Commoner, B., Corr, M., and Stamler, P. J., "The Causes of Pollution," *Environment*, 13, no. 3 (April 1971), pp.
4. Tsukui, J., "Turnpike Theorem in a Generalized Dynamic Input-Output System," *Econometrica*, 34, no. 2 (1966), 396–407.
5. Murakami, Y., Tokoyama, K., and Tsukui, J., "Efficient Paths of Accumulation and the Turnpike of the Japanese Economy," in A. P. Carter and A. Brody, eds., *Contributions to Input-Output Analysis*, Vol. 2 (Amsterdam: North Holland Publishing Company, 1970), 24–47.
6. Brody, Andrew, *Proportions, Prices and Planning* (Amsterdam: North-Holland Publishing Company, 1970).
7. U.S. Department of Labor, Bureau of Labor Statistics, *Projections 1970—Interindustry Relationships—Potential Demand Employment*, Bulletin 1536 (Washington: Government Printing Office, 1966).
8. Fisher, W. H., and Chilton, C., *An Ex Ante Capital Matrix for the United States, 1970-1975* (Columbus: Battelle Memorial Institute, March 1971).
9. See below, Chapter 2, Rudyard Istvan, "The Environmental Impacts of Electric Power Production."
10. Just, James E., "Impacts of New Energy Technology Using Generalized Input-Output Analysis" (Ph.D. dissertation, Cambridge: M.I.T., 1972).
11. Federal Power Commission, Bureau of Power, *Trends and Growth Projections of the Electric Power Industry* (Washington: Government Printing Office, 1969).
12. Leontief, Wassily W., "Environmental Repercussions and the Economic

Structure: An Input-Output Approach," *Review of Economics and Statistics*, 52 (1970), 262–271.

13. Ayres, Robert U., and Gutmanis, Ivars, "Technological Change, Pollution and Treatment Cost Coefficients in Input-Output Analysis," in Ronald G. Ridker, ed., *Population, Resources, and the Environment*, Vol. 3, *The Commission on Population Growth and the American Future: Research Reports* (Washington: Government Printing Office, 1972).

14. Leontief, Wassily W., "The Dynamic Inverse," in A. P. Carter and A. Brody, eds., *Contributions to Input-Output Analysis*, Vol. 1 (Amsterdam: North-Holland Publishing Company, 1970), 17–46.

15. See below, Chapter 3, Rudyard Istvan, "Interindustry Impacts of Projected Electric Utility Capital Formation."

16. Solow, Robert M., "Technical Change and the Aggregate Production Function," *Review of Economics and Statistics*, 39 (1957), 312–320.

17. Nordhaus, William D., "The Recent Productivity Slowdown," *Brookings Papers on Economic Activity*, Vol. 3 (1972).

18. Almon, Clopper, Jr.: Buckler, Margaret B.; Horwitz, Lawrence M.; and Reimbold, Thomas C., *1985 Interindustry Forecasts of the American Economy*, INFORUM Research Report No. 9 (College Park, Md.: University of Maryland Bureau of Business and Economic Research, August 1973) (preliminary).

I

Changing Energy

Technologies

2

The Environmental
Impacts of Electric
Power Production

RUDYARD ISTVAN

1.0 INTRODUCTION

Continuing economic growth in the United States has given rise
to a host of problems, probably few as controversial as pollution.
The signs of pollution are easy to recognize: closed beaches,
smoldering dumps, belching smokestacks. As might be expected,
electric utilities are among the principal targets of environ-
mentalists. Since they are enormous fixed-point consumers of
fuel, electric generating plants are prime pollutors. For the same
reason, they can also support intensive pollution abatement
technologies that are economically feasible. This chapter will
review (1) estimates of pollutants produced by the electric
utility industry, (2) abatement strategies that can be used to
diminish or eradicate pollutants, (3) the cost of these strategies,
and (4) their probable impact on the price of the industry's
homogeneous good, electricity. Although the estimates of impacts on the price of electricity may not be precise, the order of
magnitude will suggest which abatement strategies are possible.

The utility industry produces a common range of pollution
problems. The major problem, air pollution, results from the

consumption of fuels to fire the boilers of steam turbine generator units. A second problem, thermal pollution, occurs because
approximately 60 percent of combustion heat is wasted. Almost
one half of all water use in the United States is for cooling and
condensing purposes, and the utility industry uses 80 percent of
this volume [20]. Although there are no significant liquid wastes
produced by the industry, there is at least one solid material—
radioactive waste from fission reactors ("radwaste"), potentially
the most dangerous pollutant the economy produces. The United
States is not likely to relinquish the benefits that electricity
confers in order to eliminate the pollution that it causes, but the
industry will be forced to internalize the costs of its pollution.

The following sections of this chapter review the probable
costs of abatement for each of the pollutant groups mentioned
above and the impact of abatement strategies on the price of
electricity. Cost ranges are estimated for each assumption about
the degree of pollutant control that may be required of the
industry by the federal government. Capital and operating costs
for the most economical abatement strategies are developed, as
well as their cumulative impact on the price of electricity. In
the final section of the chapter, the results are summarized and
recommendations are made regarding the utilities' internaliza-
tion of these costs.

2.0 AIR POLLUTION

Air pollution is the focus of much of the concern about environmental problems caused by the utilities. It is a heterogeneous
category, comprised of at least five major pollutants: fly ash,
sulfur oxides, nitrogen oxides, hydrocarbons, and carbon monoxide. The first, fly ash (soot), has always been a hazard in coal
combustion. Since utilities burn coal in a powdered form (pulverized coal), about 80 percent of the ash content of the fuel is
converted to fly ash. Other air pollutants are gaseous and invisible.

Sulfur oxides cause direct biological damage. Sulfurous pollutants can be catalyzed in the presence of sunlight into sulfur
trioxide (SO_3), which turns into sulfuric acid on contact with
water. Nitrogen oxides, a third group, decompose in the presence
of sunlight to yield free atomic oxygen and nitrogen monoxide.
Atomic oxygen combines with oxygen molecules to form ozone,
a lung irritant and corrosive agent. Nitrogen monoxide recom-
bines photochemically with hydrocarbon pollutants, the fourth
type, to form compounds like ethyl alcohols and formaldehydes,
constituents of photochemical smog. Hydrocarbons, especially
polynuclear hydrocarbons, are carcinogenic. In electric utilities
most unburned hydrocarbons are entrained with fly ash. A fifth
pollutant, carbon monoxide, has well-known poisonous effects.
The gas is released when combustion is incomplete or inefficient.
Relatively little carbon monoxide is produced by the electric
utilities, but the industry produces large quantities of carbon
dioxide. Whether one considers carbon dioxide a pollutant depends on the relative importance given to the "greenhouse
effect." In this report, carbon dioxide is not considered a pollutant.

Utilities produce the major share of only one of the pollutants listed above, sulfur oxides. Automobiles alone account for
about 60 percent of all air pollution. Table 1 shows the utilities'
contribution to air pollution in tons and percentages of sulfur
oxides, nitrogen oxides, and particulates. Were it not for existing
federal controls, the figure for particulate emissions would be
even higher, perhaps 30–39 million tons in all.

It should be noted that one strategy for abating air pollutants
is the construction of nuclear generating capacity. When consider-
ing the overall price impact of air pollution abatement, nuclear
power will be discussed in this report. In the sections that immediately follow, however, only the costs of air pollution abatement for existing and planned fossil-fuel generators are considered.

Since 1971, the utilities have been required to reduce emissions from all new plants. Standards for (1) stationary fossil-fuel-

Table 1. *Emissions from Coal-Fired and*
Oil-Fired Steam-Electric Power Plants
(millions of tons/year)

Source	Sulfur Oxides Amt.	% of U.S. Total	Nitrogen Oxides Amt.	% of U.S. Total	Particulates Amt.	% of U.S. Total
Coal-fired	13.0	42.7	2.7	15.8	2.26	10.5
Oil-fired	1.0	3.3	0.3	1.8	.03	0.1
Natural gas	*		0.5	2.9	*	
Total	14.0	46.0	3.5	10.5	2.29	10.6

*Negligible
Source: N.A.P.C.A. special background material for the Joint Committee
on Atomic Energy, 1969 [33].

fired boilers exceeding 250 million B.T.U. per hour heat input,
and (2) beginning construction or modification after August 17,
1971, were published in the *Federal Register* of December 23,
1971. So far, these strict federal standards have only been imposed on new plants. It is likely that they represent maximum
standards for at least the next decade. More stringent controls
would be demanded if these standards were also applied to existing plants, and backfitting of abatement equipment would cause
changes in operating modes. This is not likely to occur. Minimum restrictions would be imposed if the 1971 standards were
relaxed. These 1971 standards serve as norms for gauging air
pollution costs and price impacts.

2.1 Particulates

Particulates are a major problem for coal plants. In a modern
pulverized coal system, incoming coal is ground to a consistency
between sugar and talcum powder, and then blown into the
boiler suspended in the combustion air supply. Of the 322 mil-

lion tons of coal consumed by the utilities in 1970, bituminous (soft) coal accounted for 316.5 million tons. It is far from homogeneous, and ranges in ash content from about 2 to 20 percent by weight [16]. Jenkins [25], reviewing various estimates of average ash content in utility coal [34, 35, 36] concluded that the national average is about 10.5 percent, or roughly 34 million tons of ash at 1970 coal consumption levels. In a typical dry bottom pulverized coal boiler, about 80 percent of this ash (27 million tons) is entrained with stack gases as fly ash; the remainder stays in the furnace as boiler clinker and slag. The National Ash Association estimates that by 1980 the total production of ash could reach 50 million tons per year. The Federal Power Commission [19] estimates that 36 percent of all 1980 generation of electricity will be from coal-fired steam plants. At present the approximate national average of coal consumption is 0.88 lbs. per kwhr. Projections indicate that there will be 2787 billion kwhr of electricity generated in 1980, which would mean 441 million tons of coal consumption and 46 million tons of ash—a confirmation of the Federal Power Commission's estimate.

For both old and new coal plants four basic types of particulate-control equipment are available: electrostatic precipitators, mechanical collectors, fabric filters, and wet scrubbers. Precipitators are the most widely used, especially in new plants. As stack gases pass through the precipitator, suspended particles of fly ash are charged with high voltage static electricity. These particles then pass between plates of opposite polarity and are attracted out of the gas to the plates. (The efficiency and cost of precipitators are affected by control strategies for sulfur oxides, as will be explained below, Section 2.2.) Mechanical collectors are usually of the cyclone type, using centrifugal force to cleanse the stack gas. They are unable to attain efficiencies above the 90–95 percent range and so must be used in conjunction with other devices (e.g., precipitators) to meet strict federal standards. Few are installed in new plants [8]. Wet scrubbers trap particulates in a fine liquid spray. There is one large disadvantage to this method. Since wet scrubbers cool stack gases, which must then be reheated in order to rise up the stack, there is a significant waste of heat involved. This means that plant efficiency is lowered and costs increased. At the present time wet scrubbers are seldom used. If they can be refined to remove sulfur oxides as well as particulates, however, their use may increase in the future. And if future regulations require removal of particles of very small size (less than 0.05 microns), wet scrubbers might prove a necessary supplement to precipitators [15]. The fourth control method involves the use of fabric filters to trap particles. These are usually arrayed in large "baghouses" to process the enormous volume of effluent. Problems with conventional fabric filters include acid destruction and pressure loss in the stack gas. Although fabric filters are seldom used for particulate control, baghouses are being installed on one plant in the western United States, with two 770 mwe units scheduled to go on line in 1973.

Costs of the commonly used precipitators for new capacity are fairly well known in the utility industry. A rough rule of thumb is that precipitator costs will be 3–5 percent of the installed cost of new capacity, depending on the required efficiency. For a plant of 500–800 mwe capacity, a precipitator with design efficiency of 95 percent, treating 300° F. flue gas and handling easily collected fly ash, has an installed cost of $1400–2200 per mwe [21]. Consolidated Edison Company of New York estimated in March 1967 that it would cost $3330 per mwe to install a 95 percent efficient precipitator on a 1000 mwe unit burning pulverized coal. For the same unit, a precipitator with 99 percent efficiency would cost $5200 per mwe [14]. Similar figures were used in a more general study [11].

Perhaps the most persuasive estimates are by Jenkins [24], who used engineering data from reference [36] to develop aggregate investment costs for particulate control. His estimate of precipitator investment per mwe of capacity is $1759 for 1969. This is consistent with the figures cited above, given unit size considerations. Jenkins' figure explicitly takes into account the higher costs of efficiencies required by the new emissions regula-

Table 2. *Costs of Backfitting*

Age of Unit in Years	Total	Total Megawatts	Dollars per Kilowatt
19	$1,715,000	207	$ 8.30
20	1,502,000	151.6	9.92
30	1,700,000	76.6	22.20
29	2,710,000	113.3	23.90
—	3,700,000	172	21.51
—	1,125,000	66	17.05
27	2,346,000	195	12.00
17	3,970,000	430	9.25
16	11,700,000	880	13.30
8	1,137,000	265	* 4.30

*The cost was very low in this case because the old unit could share a stack with a unit just being installed.
Source: Edison Electric Institute.

tions, and we can accept it as a best estimate for particulate control costs for new plants.

Maximum particulate abatement costs would be incurred if backfitting old plants were to become necessary. The cost of backfitting precipitators to existing plants can be high in relation to the future net worth of a plant. In some cases it might be more sensible to close an old plant than to backfit it with expensive new equipment. Table 2 shows actual costs of several backfitted installations. (Data were supplied by the Edison Electric Institute.) Backfitting costs were about 10 percent of the original plant costs in this age and size class. To apply figures like those in Table 2, it is necessary to estimate the amount of capacity that requires backfitting. Two studies are available. A 1969 report estimated that on a collective basis, 86 percent of all utilities' particulate emissions were controlled as required [33]. A 1970 report estimated that only 80 percent of emissions were satisfactorily controlled [13]. At most, then, about 20 percent of 1970 coal capacity would require backfitting.

For each precipitator there are three types of operating cost. Some portion of the plant's power output must be used to run the collection system. This load may be as much as 3 percent of total plant output but is more commonly about one percent. Maintenance expenses, the second type of cost, are comparatively insignificant. A third set of costs is related to the disposal of collected fly ash. Since it is difficult to handle, fly ash is usually mixed with water to form a slurry, which is then pumped or tanked to a disposal site.

A unique aspect of ash is that it need not be a waste material. It is a lightweight, chemically stable substance that can be used as a filler material in concrete, for earth stabilization, or as part of the composition of other construction aggregates. Currently, only about 20 percent of all fly ash is being utilized economically; the rest is dumped [12]. In recent years, however, the trend toward ash usage has been accelerating. Usage in 1959 was only about one million tons; in 1968 it was five million tons, with the greatest increase coming between 1967 and 1968 [40]. The costs of ash disposal systems have been estimated at $1.5/kw of capacity [14], or about $2.00 per ton of the material produced. This includes costs for the storage site and the transportation system. In the future disposal costs could be reduced to about $1.00–1.50 per ton as more of the material is sold as a by-product.

These cost estimates accumulate directly. To calculate maximum costs, we assume backfitting on all old plants to bring their collection efficiency to the level of the federal standards for new plants. Estimates of probable costs assume no backfittings. Estimates of minimum costs do not allow for reduced abatement costs resulting from increased use of low sulfur coals. Table 3 shows these estimated costs by category. Because the utilities already control particulates to some degree, as much as 80 percent of the costs are presently included in electricity prices. The generation mix by fuel type is that estimated by the Federal Power Commission.

Table 3. *Estimated Costs of Ash Collection (mills/kwhr)*[a]

Source	Maximum	Probable	Minimum
Amortization[b]			
new plant (.05 (80%))	.058	.050	.040
old plant (.09 (20%))		—	—
Disposal[c]	.090	.090	.050
Operating costs	.050	.050	.040
Total impact	.198	.190	.130
already absorbed 80%	.158	.151	.104
To be absorbed	.040	.038	.026

[a] Average yearly generation in Kwh of installed capacity = 4900; plant load factor = 56%.
[b] Straight line depreciation assumed on total plant costs of $165/kwh with 35-year life.
[c] At $2/ton of ash, .88 lb. coal/kwhr, 10.5% ash content; minimum cost with byproduct credit of $1.00/ton.

2.2 Sulfur Oxides

Sulfur oxides are far more difficult to control than particulates. Sulfur exists in both coal and residual oil fuels as a trace impurity, but the problem is particularly acute with coal. The 1970 sulfur content in coal for new steam stations averaged about 3 percent [15, 18]. About three quarters of the estimated remaining (unmined) coal reserves contain less than one percent sulfur. Most of these are subbituminous and lignite coals located west of the Mississippi River [6], far from principal load centers and unsuited to present boiler construction.

Sulfur content in coal found in the United States ranges from about 0.5 to 6 percent. The sulfur is present in two forms: 40 to 80 percent is bound as organic compounds, and the rest is in im-

bedded grains of pyrites and sulfates [39]. Hence, simple
expedients like crushing and working have limited utility in
reducing most high sulfur content coals to levels that will meet
emissions standards. Virtually all the sulfur present in the fuel is
converted to sulfur oxide pollutants when burned. There are two
possible control strategies: fuel switching (to low-sulfur coal or
to oil) and some form of stack gas scrubbing system.

Several economic factors influence the feasibility of fuel
switching. The majority of low sulfur coals are located far from
eastern markets, where the need for them is greatest. Transportation costs may render a switch to low-sulfur coal more costly
than other control techniques. Opening new mines is a substantial undertaking. A typical mine today requires an investment of $10–12 per annual ton of capacity, of which 75 percent
is for mining equipment. As a result, most mining companies
require a firm commitment from users before opening a mine.
For new generating stations, this usually involves a thirty-year
contract for more than one million tons per year. Letting such
contracts for old plants with short lifetimes is generally not
feasible, although it may be possible for new plants.

Perhaps the single most important consideration inhibiting
the use of low-sulfur coal is that deposits within reasonable
distance of markets are actually more valuable as metallurgical
coking coal than as boiler fuel for utilities. Most low-sulfur coals
east of the Mississippi are found in deep, thin seams in West
Virginia. Despite high mining costs, steel manufacturers mine
these seams because they can realize major cost savings. As a
general rule, a 0.1 percent reduction in the sulfur content of
metallurgical grade coal is worth about $.50 per ton of coal to
steel companies. These coal reserves are almost completely
owned by U.S. steel companies or are under contractual commitment to them for their own use or for export to foreign
steelmakers.

Some of the costs involved in switching from high- to low-
sulfur coal are due to the nature of the coal itself. Since each
boiler is designed to handle a specific coal, before a different

coal can be burned, costly modifications may be necessary. Even with such modifications there may be a loss of thermal efficiency. The ash fusion temperature, in particular, must be higher as sulfur content declines. In addition, sulfur oxides act as a conditioning agent for fly ash, enabling it to accept an electrical charge. For low-sulfur coal there is a higher resistivity of fly ash. Thus collection costs rise. For a cold-side precipitator (used for boilers whose flue gas temperature is about 300° F.), costs are higher when low-sulfur coal is burned. One solution has been to install, or backfit, hot-side precipitators, operating at temperatures of about 700° F. This can mean precipitator investment costs two and a half to three times higher than those cited in Table 1 [38].

Switching from coal to oil is also technologically possible. It requires major plant modifications in boiler design and fuel handling facilities. Some new plants are being designed with the ability to burn either fuel. Although much of the residual oil, particularly that from Venezuela, is high in sulfur, it can be desulfurized. Most major refiners have already added, or are in the process of adding, hydrodesulfurization facilities in their plants for desulfurizing Caribbean oils. Bechtel Corporation [38] found that lowering residual oil to 0.5 percent sulfur would require incremental investments in the range of $100–150 million per plant. Costs per barrel of processed residual range from $.24 to 1.02/bbl on a five-year payout basis, depending on the degree of desulfurization required and the processing method.

Despite the substantial fuel premium, many utilities are finding that they must use desulfurized residual oil to meet present emissions standards for existing plants in metropolitan areas. Low sulfur coal is simply not available. The final price premium will depend in part on oil import quotas and world oil demand as well as desulfurization costs. Most Middle Eastern crude is naturally "sweet," producing undesulfurized residual oils containing less than one percent sulfur. But for political reasons utilities may not be able to avail themselves of this oil.

Because of the costs of fuel switching, other solutions have

Table 4. *Comparative Costs of Sulfur Dioxide Removal Processes*

Process	Total cost ($/ton of coal) (when no credit is taken for sale of by-product)	(when credit is taken for sale of by-product)	Source of credits
Reinluft [3]	2.45	1.30	concentrated acid at $23.50/ton
Alkalized alumina [26]	1.54	0.86	sulfur at $25/ton
		0.32	sulfur at $45/ton
Catalytic oxidation [27]	1.75	0.72	moderately concentrated acid at $9/ton
		0.38	acid at $12/ton
DAP-Mn [1]	4.39[a]	1.10[a]	ammonium sulfate at $32/ton
Kiyoura TIT [28]	3.66[a]	0.44[b]	ammonium sulfate at $32.20/ton
Dolomite (Combustion Engineering, Inc.) [41]	0.36–0.63[b]	0.36–0.63[b]	
Wellman-Lord, Inc. [42]	0.73	0	sulfur at $25/ton
		(–) 0.60	sulfur at $45/ton

[a]Per metric ton of fuel oil.
[b]There is no salable by-product from the dolomite process, but in installing this process, certain capital equipment savings and operating savings (estimates at $.27/ton) are expected.

been sought. Two common relief measures have been mine-mouth plant siting and the use of taller stacks. With the former, pollution is simply located elsewhere. Neither method meets emissions standards for new stations. The major alternative to fuel switching is some form of stack gas scrubbing system. Over 30 such systems have been proposed. One—Combustion Engineering's limestone-dolomite injection system—has been installed on

Table 5. *Cost of the Cat-Ox System*

	Addition of Cat-Ox to 500 mw., existing powerplant	Installation of Cat-Ox with new 1,000 mw. powerplant
Capital cost	$18,000,000	$32,300,000
Cost per kilowatt	$36	$32.30
Added operating cost, in cents per million btu (gross)	6.5	3.3
Acid credit	2.3	2.4
Net	4.2	.9
Fixed costs (at 16%) in cents per million btu	8.7	8.2
Total costs, in cents per million btu	12.9	9.1
Total costs, in mills per kilowatt-hour	1.23	.82

Source: Monsanto Enviro-Chem Systems, Inc.

commercial units on an experimental basis. In this simple system, limestone or a similar mineral is pulverized and injected with pulverized coal into the boiler, where it absorbs sulfur chemically. The system has been found to create more than twice the normal amount of fly ash (effectively putting precipitators out of action). In addition, it fouls boiler walls. The test unit has been taken out of service for an overhaul [32]. Perhaps with further refining, this or a similar system will reduce sulfur oxide emissions and will cost less than a fuel-switching strategy. Table 4 lists some processes that may solve the problem, with their estimated cost per ton of coal. The by-product credits shown in the table are probably optimistic, however. Should any such process be put into common use, the volume of sulfur by-products would be hundreds of thousands of tons per year at current coal consumption rates; by-product markets would deteriorate.

To establish more precise estimates of the costs involved in installing such a system, we broke down the costs for the catalytic oxidation system made by Monsanto Enviro-Chem Systems, Inc.

Table 6. *Sulfur Oxide Abatement Costs
(mills/kwhr)*

Source	Maximum	Probable	Minimum
Coal[a]	$ .88	$ .66	$ .27
Oil[b]	1.26	.85	.30
Average cost at 1970 mix	.588	.44	.17

[a] at 0.88 lb./kwhr.
[b] at .0017 bbl./kwhr.

(Table 5). Even for a new plant, the total estimated cost per kwh
is substantial: roughly comparable to a $1.90-per-ton fuel pre-
mium. In any event, stack gas scrubbing systems will not be
available before 1975—possibly not before 1980. In extreme
cases low sulfur fuel premiums may exceed scrubbing system
costs. In other words, the cost of sulfur oxides abatement is the
largest direct environmental cost facing utilities. Under the feasible fuel switching strategy, we assume $2.00 per ton of coal and
$.75 per barrel of oil premiums and no interfuel substitution for
the maximum case where new standards are applied to all plants.
Compared to these operating charges, conversion costs are relatively insignificant. More probable is application of observed
1970 premiums of $1.50 per ton coal and $.50 per barrel of oil
to new plants with limited fuel switching; minimum cost assumptions include $1.25 per ton of coal and $.35 per barrel of oil premiums on only 50 percent of new plant (that in urban areas).
Table 6 shows these results, using plant fuel mix and new plant
estimates detailed in the footnote to Table 10.

2.3 Nitrogen Oxides

The final air pollutant of concern to electric utilities is the
nitrogen oxides group. These pollutants are formed at high
temperatures in most combustion processes. They decompose

rapidly at combustion temperatures. Only the nitrogen oxides that reach the superheaters at the top of a boiler will enter the atmosphere [43]. Nitrogen oxides will not form unless excess oxygen is present in the boiler. Thus the simplest control method would seem to be closely controlled combustion stoichiometry. The Edison Electric Institute indicates, however, that the stoichiometric regulation of combustion—or even modification of boiler design to achieve burner combinations less favorable to nitrogen oxide formation—would have no significant effect on this pollutant. On the other hand, the Environmental Protection Agency reported after a series of tests that over 80 percent of the test boilers could be successfully modified to meet nitrogen oxides emission standards [10]. No costs were reported for such conversions.

Nitrogen oxide emissions do not arouse the same public concern as do sulfur oxide pollutants. It is therefore unlikely that existing capacity will be backfitted with combustion modifications to control nitrogen oxide levels. This is particularly so because automobiles rather than utilities are the major source of nitrogen oxide emissions. The cost of design modifications on new plants to meet the new emissions levels is insignificant compared to the costs incurred from other sources. For practical purposes, the price impact of nitrogen oxide control can be considered negligible.

3.0 WATER POLLUTION

The major water pollutant produced by electric utilities is heat. One of the unalterable characteristics of thermodynamic processes is limited efficiency. Not all the heat energy produced in thermal generation is converted into electricity; some results in unconsumed (waste) heat. The theoretical efficiency of the steam turbine (Rankine) cycle is about 60 percent, but because of design limitations the present maximum attainable real efficiency is about 40 percent. The amount of waste heat produced at this efficiency will be on the order of 60 percent or more of all the energy consumed in power generation.

Water in some form is the most commonly used heat sink. It is an efficient coolant for most industrial processes. Water dissipates its heat load to the atmosphere through evaporation. It can be viewed as a convenient transport mechanism rather than an ultimate heat sink.

Elevations in water temperature due to waste heat removal has been appropriately termed "thermal pollution." The effects of thermal pollution fall into three categories: economic, physical, and biological.

The economic effects are twofold. Cooling water is a consumptive use. It was noted above that the atmosphere serves as an ultimate heat sink. Bodies of water heated above ambient temperature levels yield this heat to the atmosphere, primarily via evaporation but also through other mechanisms. The evaporation of one pound of water requires about 1050 btu, or 8762 btu per gallon of water evaporated. In a hypothetical 1000 mwe station with 35 percent efficiency, this amounts to an evaporative consumption of about 410,000 gallons of water per hour of operation if the water returns to its ambient temperature strictly through evaporation. The second economic effect is more subtle. If water is to be used repeatedly for cooling—as, for example, at a succession of plants along a river—the temperature of the water at the intake is important, for turbines become less efficient as the cooling water temperature increases because of the shallower temperature gradient. This effect could cause significant costs for downstream plants. Lof and Ward, calculating the economic impact of thermal pollution from steam electric plants on subsequent cooling water uses, concluded that there are modest increases in operating costs by downstream users [31].

Water is also changed physically by thermal pollution. The viscosity of the water decreases, hastening evaporation and flow rates. Because heat reduces density, thermally polluted water "floats" on the surface. It can form a thermal boundary or

contribute substantially to temperature stratifications.* Perhaps
the most important physical effect, however, is the increase in
dissolved mineral concentrations in the water due to evaporative
loss.

Through changes in mineral concentrations, thermal pollution
also affects the aquatic biosphere. Most of the arguments commonly heard about thermal pollution center on this third aspect
of thermal pollution: its biological effects. These are so numerous
and so subtle that simply to catalogue them is a major task.
Thermal discharges are not necessarily detrimental biologically.
For example, an experiment is under way in Maine to use power
plant cooling effluent to promote lobster production [17].
Severe temperature rises do, however, have major detrimental
impacts on the aquatic environment, and as a result, in response
to the 1966 Clean Water Act, many states have recently passed
statutory limitations on permissible degree rises in the cooling
water effluent, a typical value being a 5° F. maximum rise.† In
the future, to solve the problem, greater volumes of water will be
run through condensers to limit temperature rises, or alternative
cooling means will be employed.

Three kinds of cooling techniques are available to the utilities.
The one commonly employed is once-through use of water.
Since more than 90 percent of current U.S. capacity is already
using this method, its immediate costs are already internalized.
In the future, this method may not be sufficient to meet new
regulations governing the permissible rise in temperature. A
second method is the use of a cooling pond to retain the heated
water. Effluent is pumped into the pond and, after cooling, is
either released back to the main source or recycled to the condenser. An average pond requires one to two acres of surface
area per megawatt of capacity, depending on local climactic
conditions. The cooling process can be enhanced by spray

---

*This may be particularly important where epilimnion, thermocline, and
hypolimnion zones do not form naturally.
†See reference 17 for a complete listing of regulations.

systems that reduce the required surface area by a factor of 20 (because of the increased efficiency of evaporative cooling [20]). About 31 percent of all new plants use cooling ponds.

For a typical 1000 mwe installation, a spray pond of about 1000 acres is possible. But the number of sites that could accommodate a 2000-acre cooling pond is limited, even assuming the pond to be entirely contained in artificial dikes. In rural areas, land costs are lower and cooling ponds have been used. At Coffeen, Illinois, for example, a cooling reservoir was created at a cost of about $2.50/kw. Forty percent of this cost was for land; 40 percent, for cleaning of the lake area and construction of a dam; and 20 percent, for other development [28]. The $2.5/kw cost may have been atypically low because the site's natural contours obviated the need for extensive impoundments. Costs per kw of capacity have been estimated to range between four and six dollars for fossil-fueled steam stations and between six and nine dollars for nuclear plants of 1200–2000 mw capacity.

The third method is the cooling tower, now used by about 36 percent of all new stations. Although there are many variations in design, there are two basic types, wet and dry. Wet towers put coolant water in direct contact with the air; evaporation removes the thermal energy. Dry towers circulate coolant through a system of heat exchangers and rely on conduction and convection to transfer the thermal energy to the atmosphere (like an automobile radiator).

Wet and dry towers may be classified as mechanical draft or natural draft towers. And, in turn, mechanical draft towers may be classified as forced mechanical or induced mechanical. Both rely on fans to move air past the water to be cooled. In forced towers, fans located at the base of the unit blow air up through the tower. In induced towers, fans at the top of the unit suck air through the tower from below. There are several types of natural draft towers as well, but all of them rely on natural air circulation that is caused by the "chimney effect"—the phenomenon that causes hot air to rise in an ordinary flue.

Wet and dry towers each have good and bad points. Wet

towers are relatively more efficient than dry because they use air at the wet bulb temperature. Except when humidity is 100 percent, the wet bulb temperature will be lower than the dry bulb temperature and will produce a steeper thermal gradient. In all cooling towers there may be a loss in plant efficiency because of higher turbine backpressures from low-efficiency condensers. This may be as high as 6–8 percent in dry towers.

Wet mechanical towers are the least capital intensive of all cooling towers. Because of pressure considerations the forced draft variety are limited to fan cells of about 12 feet in diameter. The induced draft type have cells that measure up to 60 feet in diameter. In both forced and induced draft wet mechanical towers, a bank of cells is used to cool the plant condenser water. More cells are needed in the forced draft tower because of its size limitations, and this means increased capital costs.

Fans in wet induced draft towers are in contact with the moist air after it has cooled the water. The corrosion that results leads to higher maintenance costs. In either type of wet mechanical tower, forced or induced draft, water is trickled through a "packing" where it contacts the air before being collected for recirculation. Droplets of moisture are entrained in the air draft during this process. This "windage" is a major problem. Together with the evaporated water, windage may cause fogging, icing, or rain in the area of the tower (any of which may persist for up to a quarter of a mile in open areas). Since water used for cooling is treated chemically to limit build-ups of organic matter and solids in the condensers, deposition of these chemicals in the area surrounding a tower may also cause a problem. In addition, the water consumption is high.

Capital costs for both forced and induced draft wet towers are about five to eight dollars per kw. Requirements for pumps and fans may be about 0.8 percent (or more) of total plant electrical output [29]. Annual operating and maintenance costs, exclusive of power requirements, are estimated at $60,000 per year, or about one percent of the initial investment [5]. At the 540 mwe nuclear plant proposed for Vernon, Vermont, the annual production cost of power to operate the cooling unit is estimated at $160,000, or about 2 percent of the initial tower cost of $6 million [9].

Wet natural draft towers have higher initial costs but lower operating costs than wet mechanical draft towers because of the greatly reduced power requirements. The basic structure is a huge hyperbolic concrete shell, up to 350 feet across at the base and about 400 feet high. This acts as a chimney to create the draft from waste heated air. As in all wet tower designs, water is trickled or sprayed through a fill at the base of the tower while air passes through it. Although evaporative losses are present, windage is not a problem. Water losses are on the order of 2.8 percent of circulation [28]. Effects on the local climate are minimal because of the height of the tower and volume of air, although a visible plume may extend for several hundred feet. Of greater concern is the tower's aesthetic effect—because of its size it is bound to dominate its surroundings.

Typical capital costs for a natural draft wet tower are about $11/kw. Table 7 sets forth estimated costs at the TVA Paradise Plant (coal fired, 1400 mw in two units). The operating costs during the plant life have been converted to present worth values. Assuming plant life of thirty-five years (the norm), this yields a rough total yearly cost, including depreciation, of about $540,000. The figure is low because of the discounted future worth of operating costs.

Dry towers, either mechanical or natural draft, are for several reasons relatively more expensive than wet towers. First, they are less efficient because they are limited by dry bulb temperatures. Second, heat transfer is by conduction and conversion, a slower process than evaporation. (One advantage, however, is the elimination of evaporative losses.) As a result, dry mechanical and natural draft towers must be larger than comparable wet towers, with related increases in costs. Investment costs for dry natural draft towers are estimated at $25/kw; for dry induced draft towers, $27/kw. With these figures, it is natural to assume that wet towers will be built wherever possible.

Table 7. *Hyperbolic Cooling Tower Costs*
TVA Paradise Plant

Capital Costs:
1. Two cooling towers	$ 6,600,000
2. Warmwater tunnels	1,100,000
3. Cooled water flumes, etc.	1,350,000
4. Additional pumps	1,000,000
5. Underwater dam and skimmer wall	100,000
	10,150,000
General expenses, overheads, and contingencies	3,180,000
Subtotal	13,330,000

Operating Costs:[a]
1. Heat rate losses for units 1 and 2	820,000
2. Additional pumping power, units 1 and 2	650,000
3. Heat rate loss for unit 3	3,000,000
4. Additional pumping power, unit 3	1,000,000
Subtotal	5,470,000
Total for Paradise	18,800,000

[a]Additional operating costs during life of plant converted to present worth value.
Source: Tennessee Valley Authority.

To pull together economic data from the preceding analysis, Table 8 presents a comparison of various cost estimates. On the basis of Table 8 information, the cost impact of alternative cooling systems other than once-through cooling are computable. Table 9 uses only the maximum cost estimates for each alternative, and assumes a thirty-year plant life, straight-line depreciation and 60 percent utilization rates over the thirty-year period. Operating costs were assumed to be greatest for forced draft towers (5 percent of total capital cost per year), less for natural

Table 8. *Capital Cost in $/kw of Capacity for Cooling Water Systems for Steam Electric Plants*

Type of System	Korflat [29]	Perry [40]	E.E.I. [8]	E.W. [7]a	F.P.C. [17]	L&W [32]
Once-through						
fossil	—	2–3	—	—	2–3	5–6
nuclear	—	3–5	—	—	3–5	
Cooling pond						
fossil	2.5–5	4–6	—	3–5	4–6	10
nuclear		6–9	—	4–6.5	6–9	
WET COOLING TOWER						
Mechanical draft						
fossil	7	5–8	4.5–9	5–8	5–8	—
nuclear		8–11		6.5–10.5	8–11	—
Natural draft						
fossil	11	6–9	6.5–11	10–15	6–9	7.5–11
nuclear		9–11		13–19	9–13	
DRY COOLING TOWER		b				
Mechanical draft						
fossil	25	—	25–28	—	25–28	—
nuclear		—		15–20		—
Natural draft						
fossil	27	—	27–30	—	27–30	22
nuclear		—		25–50		

aIt should be noted that some other estimates in this *Electrical World* article (January 19, 1970) are surely incorrect. The estimated cost of a 1000 mwe nuclear station, for example, is given as $200 million—about $89 million below the A.E.C.'s own low estimate.
bCosts stated to be "very much higher."

draft towers (2 percent), and the same for once-through cooling and cooling ponds. These operating cost estimates are conservative (i.e., high).

Table 9. *Thermal Abatement Costs (mills/kwh)*

	Capital	Operating	Total
Once-through	—	—	—
Cooling pond			
fossil	.04	—	.04
nuclear	.06	—	.06
Wet tower			
mechanical			
fossil	.05	.09	.14
nuclear	.07	.10	.17
natural			
fossil	.10	.07	.17
nuclear	.12	.09	.21
Dry tower			
mechanical			
fossil			
	.18	.27	.45
nuclear			
natural			
fossil	.19	.14	.33
nuclear	.32	.23	.55

Similar estimates are arrived at in References 2, 4, 5, 22, 30.

4.0 SOLID WASTES

Aside from the costs of fly ash disposal discussed in Section 2, the principal problem in the area of solid wastes is the cost of disposing of radwastes, the highly radioactive fission wastes produced in nuclear reactors. Typical power reactor cores contain an inventory of nuclear "ashes" many hundred times greater than those produced by a nuclear bomb. These wastes are contained in the fuel rods used in reactor cores. Strict controls are used to keep the wastes from escaping from the core. Power plant radwastes emissions are much lower than natural background radiation.

Only a small percentage of the usable energy in nuclear fuel is consumed before the fuel rods must be removed and reprocessed. There are now several reprocessing plants in the United States designed to accept more than 1000 tons of irradiated fuel each year. In these plants fuel rods undergo three types of processing. First, the fuel elements are mechanically destroyed—chopped into pieces. Next, these pieces are dissolved in a hot nitric acid bath. Finally, usable uranium and plutonium are extracted from the solution by the Purex solvent process. Recovered nuclear fuels are shipped to other plants for enrichment and refabrication into new fuel rod assemblies.

It is in the third stage that radwastes become a problem. There the residual solution, now carrying radwastes, is neutralized and concentrated for storage. The concentrated radwastes are stored in buried stainless steel tanks. The total cost of the combined processing and waste management services set by one firm, Nuclear Fuels Service, is about 0.2 mills per kwh.

The problem associated with radwastes is the cost of stored waste concentrate. The physical bulk of concentrated wastes is small—about 10 cubic feet per year from a 1000 mwe LWR station. However, the wastes must be stored safely for long periods of time: radioactivity levels do not abate to safe levels for at least 1000 years. Any small fraction of the wastes released into the environment would be lethal in the immediate area and would have far-reaching effects. Various methods of more permanent storage have been explored—for example, entraining radwastes in glass, then depositing the bottles in an abandoned Kansas salt mine. Instability of the geologic formations made this proposal impracticable.

Since no workable solution has yet been found, the costs of an ultimate storage system are not known. If the Atomic Energy Commission's estimates are correct, by 1980 we will be processing 30,000 tons of U_3O_8 each year [44]. The nuclear engineering

department at the Massachusetts Institute of Technology estimates that eventual storage costs for these volumes of radwastes
may amount to 1–2 percent of fuel reprocessing costs, or about
.015–.03 mills/kwh. Given the total cost of nuclear energy and
the variability in the plant capital costs, radwaste disposal costs
may prove to be insignificant although the problem itself is of
major importance to the industry.

5.0 PRICE IMPACTS

Previous sections of this report estimated cost impacts of pollution control by pollutant category. If utilities pass these costs on
to consumers, electricity prices will be directly increased. Utilities
earn a regulated rate of return. Increases in costs are ultimately
reflected in higher rates. Hence, it is probable that in the long run
few if any of these costs will be internalized by the industry; they
will be passed on to consumers in full.

Not all abatement costs apply to all types of generation. Table
10 estimates total cost increases by type of fuel, using information
from Tables 3, 6, and 9. In all cases, the most conservative (i.e.,
most costly) estimates were used.

These estimates, combined with the proportion of generation
expected by 1980, suggest an overall weighted average cost increase of about 0.62 mills/kwr by 1980. This is an important
figure because it suggests the probable maximum increase in the
price of electricity related to pollution abatement costs. Different
consumer classes pay different prices; in 1970 the average electricity price was 1.61 cents/kwh. Thus a probable average price
rise is 3.8 percent. This figure represents a maximum estimate. It
was calculated using large abatement costs, some of which have
already been internalized. For example, as much as 80 percent of
particulate abatement costs may already be reflected in 1970
electricity prices. (Correcting for this inclusion would yield maximum probable costs of 0.58 mills/kwh and a 3.6 percent average
price increase.)

Table 10. *Increased Costs Due to Pollution Abatement (mills/kwh)*

	Coal	Oil	Gas	Nuclear	Other
Particulate	.198	—	—	—	—
SO_x	.88	1.20	—	—	—
NO_x	—	—	—	—	—
Thermal[a]	.032	.082	.021	.148	—
Solid	—	—	—	.03	—
Total	1.11	1.28	.021	.178	—

Total average weighted increase: 0.62 mills/kwh.

[a]20% cooling ponds, 20% natural draft wet towers, 60% forced draft wet towers. Costs assumed for new plant generation only (see Reference 2). The data below are from the *1970 National Power Survey*.
[b]Plant generation and construction mixes:

	1970	1980	% New Gen.
total net generation, mwh	1391	2787	
proportions: coal	.54	.36	.25
oil	.09	.13	.65
gas	.20	.12	.17
nuclear	.03	.28	.95
other	.14	.11	—

A 3.8 percent increase in the real price of electricity seems insignificant when compared to potential increases arising from higher capital costs and higher fuel costs as gas and petroleum come into short supply. The conclusion is inescapable: these pollution cost/price impacts are relatively unimportant.

6.0 CONCLUSIONS

The ultimate price impact of abating pollutants produced by the electrical utilities amounts to a small percentage of the cost of electricity. This is particularly true if one remembers that in

areas where restrictions may make fossil steam energy more expensive, utilities are exercising the option of turning to nuclear
generation for the bulk of their new generation, minimizing the
price impact of fossil fuel costs.

Why then is the utilities industry so slow to implement the
kind of abatement procedures that have been outlined above?
The costs are relatively small and, in theory at least, it should be
possible for utilities to appeal to rate-setting agencies for price
increases corresponding to these costs. One reason is simply
institutional inertia. For many utility executives, environmental
concerns are new. In the past, the goal has been to provide all
the power the nation demanded at the lowest possible cost.
This was not just a matter of choice. In many states the utilities
are under charter from regulatory agencies to provide service
at the lowest possible cost. Technologies that do not further
such goals are discouraged. Only recently have these institutional goals been changing.

A more important deterrent to change may be the regulatory
system that determines rate structure. Rate increases are un-
popular. Lengthy public hearings must be held before a new
rate structure is accepted. Even after hearings a utility may not
receive the total increases requested. It may take several years
for a utility to accumulate operating data to document its cost
claims and to receive compensatory rate increases. This was the
situation during the 1968–71 inflation. Rate increases did not
keep pace with rising costs and utility profit margins suffered.
If utilities were to decide on total pollution abatement, they
probably would not be allowed to raise prices right away. For
the firm that wishes to maximize profits, this regulatory lag acts
as a negative incentive.

When the technology of an industry is fairly stable and no
major new cost demands are made upon it, the type of regulatory system now operating for utilities works very well. Given
the lag between cost and rate changes, any increase in efficiency
(cost minimization) can result in short-run profit margins higher
than those granted by the regulating agency. Subsequent rates

are not usually set to compensate for such "windfall" profits. Hence, a slow rate-setting mechanism operates to stimulate efficiency when conditions are stable.

The new environmental demands being placed on the utilities change the conditions in which the system operates. As a result, utilities avoid pollution abatement and its costs in order to maximize profits. The solution seems elementary. Regulatory mechanisms should be restructured to give rate increases based on the projected costs of pollution abatement. The rate increases could go into effect along with abatement measures, or compensatory rate levels could be set later to remove economic disincentives. This would also serve to inform the public of the true costs of pollution abatement as they are incurred. Debate over environmental needs might then be less obscure. It would be easier for the average consumer to evaluate the costs of clean air, cool water and safe nuclear power. The result might be more informed public debate and more responsive governmental policy with regard to rate setting and regulation.

REFERENCES

1. Atsukawa, M., et al., "Dry Process SO_2 Removal Method," *Technical Review of Mitsubishi Heavy Industries* (January 1967), pp. 33–38.
2. Battelle Memorial Institute, Pacific Northwest Laboratories, "Nuclear Power Plant Siting in the Pacific Northwest," unpublished paper prepared for the Bonneville Power Administration (1967).
3. Bienstock, D., et al., "Evaluation of Dry Processes for Removing Sulfur Dioxide from Power Plant Flue Gases," *Journal of the Air Pollution Control Association* (October 1965).
4. Christianson and Tichenor, "Economic Aspects of Thermal Pollution Control in the Electric Power Industry," *Environmental Effects of Producing Electric Power*, Part 1 (Washington: Government Printing Office, 1970).
5. Converse, A. O., "Thermal Energy Disposal Methods for the Proposed Nuclear Power Plant at Vernon," in *Thermal Pollution Hearings*, Part 1 (Washington: Government Printing Office, 1968), pp. 354–403.

6. DeCarlo, J. A., et al., *Sulfur Control of United States Coals* (Washington: Bureau of Mines, 1966).
7. Edison Electric Institute, "Cut Pollution at What Cost," *Electrical World* (January 19, 1970).
8. —— "Response to Question Put by the Joint Committee on Atomic Energy," *Environmental Effects of Producing Electric Power*, Part 2, Vol. 1 (Washington: Government Printing Office, 1970).
9. Elinka, S. M., "Cooling Towers," *Power* (March 1963).
10. Environmental Protection Agency, "Control of Air Pollution from Fossil Fuel Fired Steam Generators," unpublished study (1971).
11. Ernst and Ernst, *Costs and Economic Impacts of Air Pollution Control* (Washington: U.S. Public Health Service, October 1969).
12. Faber, J. H., "Coal's Generation Primacy Hinges on Fly Ash Sales," *Electrical World* (April 28, 1969), pp. 44–45.
13. Fogel, Johnston, et al., *Comprehensive Economic Cost Study of Air Pollution Control Costs for Selected Industries and Selected Regions*, Report R-OU-455 (North Carolina: Research Triangle Institute, February 1970).
14. Federal Power Commission, *Air Pollution and the Regulated Electric Power and Natural Gas Industries* (Washington: Federal Power Commission Office of Public Information, September 1968).
15. ——, "Eleventh Steam Station Design Survey," *Electrical World* (October 15, 1970).
16. —— "Fuels and Fuel Transport," unpublished paper.
17. —— *Problems in Disposal of Waste Heat from Steam-Electric Plants* (Washington: Federal Power Commission, Office of Public Information, 1969).
18. —— "Tenth Steam Station Design Survey," *Electrical World* (October 21, 1968).
19. —— *1970 National Power Survey* (Washington: Government Printing Office, 1972).
20. Federal Water Pollution Control Administration, *Industrial Waste Guide on Thermal Pollution* (Corvallis, Pacific Northwest Laboratory, September 1968).
21. Gallear, R., "Cost Factors in the Selection of Cyclone and Electric Precipitators for Fly Ash Collection," unpublished paper (Buell Engineering Company, Lebanon, Ohio).
22. Hauser, V. L. "Cooling Water Requirement for the Growing Generation Additions of the Electric Utility Industry" (paper presented at the American Power Conference, Chicago, April 22–24, 1969).
23. Jenkins, T., "Aggregate Investment Cost to the Electric Power Industry of Particulate Air Pollution Control Equipment," unpublished

paper (Cambridge, Massachusetts, Harvard Economic Research Project, March 1972).

24. —— "The Required Efficiency of Particulate Collection Equipment in the Electric Power Industry," unpublished paper (Cambridge, Massachusetts, Harvard Economic Research Project, January 1972).

25. Katell, S., "Removing Sulfur Dioxide from Flue Gases," *Chemical Engineering Process* (October 1966), pp. 67–73.

26. Katell, S., and Plant, K. D., "Here's What SO_2 Removal Costs," *Hydrocarbon Processing* (July 1967), pp. 161–164.

27. Kiyoura, R., "Studies of the Removal of Sulfur Dioxide from Hot Flue Gases to Prevent Air Pollution," *Journal of the Air Pollution Control Association* (September 1966).

28. Korflat, P., Testimony before the Subcommittee on Air and Water Pollution, Senate Committee on Public Works, in *Thermal Pollution Hearings,* Part 1 (Washington: Government Printing Office, 1968).

29. —— "Thermal Discharges," in *Industrial Water Engineering* (March 1968).

30. Lof and Ward, "Economic Considerations in Thermal Discharge to Streams," (paper presented at the National Symposium on Thermal Pollution (II), Nashville, Tennessee, August 14–16, 1968.

31. —— "Economics of Thermal Pollution Control," *Journal of the Water Pollution Control Federation* (December 1970).

32. Martin et al., "Air Pollution Control Systems" (paper presented at the 1970 Industrial Coal Conference, University of Kentucky, April 8–9, 1970).

33. National Air Pollution Control Administration, "Air Pollution from Steam-Electric Generating Stations," *Environmental Effects of Producing Electric Power,* Part 1 (Washington: Government Printing Office, October 1970).

34. National Coal Association, *Electrical World* (October 18, 1965).

35. —— *Steam-Electric Plant Factors* (Washington: 1968).

36. Oglesby, S., Jr., *A Manual of Electrostatic Precipitator Technology,* Part II (Birmingham, Alabama, Southern Research Institute, August 1970).

37. —— "Electrostatic Precipitators Tackle Air Pollutants," *Environmental Science and Technology* (September 1971).

38. —— *Wall Street Journal* (June 19, 1967).

39. Perry, H., "Potential for Reduction of Sulfur in Coal by Other Than Conventional Clearing Methods" (paper presented at the Symposium on Economics of Air Pollution Control, 59th National Meeting of the American Institute of Chemical Engineers, Columbus, Ohio, May 15–18, 1966).

40. —— Testimony before the Joint Committee on Atomic Energy, Novem-

ber 4, 1969, in *Environmental Effects of Producing Electric Power*, Part 1 (Washington: Government Printing Office, 1970).

41. Plumley, A. L., Whiddon, O. D., et al., "Pilot Plant Absorbs Sulfur in Station Stack Gases" (unpublished paper, Hartford, Conn.: Combustion Engineering, October 1967).

42. —— "Removal of SO_2 and Dust from Stack Gases" (paper presented at American Power Conference, April 25–27, 1967).

43. Sensenbaugh, W. O., "Formation and Control of Oxides of Nitrogen in Combustion Processes," (unpublished paper, Combustion Engineering, Inc., 1969).

44. U.S. Atomic Energy Commission, Division of Reactor Development and Technology, *Current Status and Future Potential of Light Water Reactors* (Washington: Government Printing Office, March 1968).

3
Interindustry Impacts
of Electric Power Production

RUDYARD ISTVAN

1.0 INTRODUCTION

In recent years brownouts and blackouts have come to symbol-
ize the energy crisis in electricity. With electricity consumption
increasing about 7.2 percent per year, the utility industry has
been hard-pressed to construct sufficient capacity to meet
demand. The Federal Power Commission estimates that utilities
may require as much as $450 billion of investment by 1990 [2],
but no one is sure that the industry will be able to raise these
funds. Equally uncertain is the impact that investment spending
of this magnitude will have on the economy. In this chapter,
some of these impacts will be examined on an industry-by-
industry basis.

The problem of utility investment is one of complexity as
well as magnitude. In addition to the need to double capacity
every ten years, increasing the peak load and unit sizes requires
more than proportionate increases in peaking and reserve capacity. Long lead times are involved. Indefinite orders are placed on
turbine manufacturers' books ten years in advance of delivery.
In the past, nuclear plants have taken more than seven years to

construct. These long lags between initial investment and completion of facilities produce a complex pattern of economic impacts. A general dynamic model developed in Section 2 can be used to explore the complex interrelationships among investment and production levels.

Investment problems are compounded by uncertainties about technology and cost. Rising fuel prices and increasingly expensive pollution restrictions are changing the relative merit of alternative generating technologies. Investment in formerly inexpensive coal-fired plants may be shifted to construction of residual oil or nuclear plants, causing differing impacts on the economy. Increasing voltage levels and changing plant-siting norms are altering the nature of transmission facilities. Increasing use of underground urban residential distribution has changed the pattern of distribution system requirements. The input-output matrices specified in Section 3 are designed to explore such alternative utility investment strategies. With these partitioned matrices, it is possible to examine the effects of change within technologies and, more important, change in the mix of technologies.

Specifications and data to implement the input-output model are developed in Section 3. In Section 4, the model is used to examine two sets of issues surrounding alternative utility investment strategies. First, the economic impact of projected technological changes to 1980 is explored. Such changes include the construction of pollution abatement equipment, extra high voltage transmission, underground distribution, pumped storage peaking plants, and increasing unit sizes. Second, the economic impacts of two major utility investment strategies are explored—fossil-fuel generation (with appropriate abatement equipment), and nuclear generation.

The principal finding is that changes in fossil-fuel technology affect some, but not many, industries. On the other hand, construction of nuclear plants places much more diverse strains on the economy. Almost all sectors are affected to some degree.

2.0 A GENERAL DYNAMIC INPUT-OUTPUT MODEL

The lagged dynamic input-output model developed for this study is an extension of the Leontief dynamic model [7]. In the Leontief formulation, the fundamental input-output accounting identity is modified to separate investment from other final demands with an explicit term:

$$X_t = A_t X_t + K_t + C_t \tag{1}$$

where A_t is the technical coefficient matrix for period t
K_t is a vector of investment for period t
C_t is the vector of consumption for period t
X_t is the vector of total output for period t.

The investment vector K_t is determined by a simple form of the accelerator model:

$$K_t = B_t (X_{t+1} - X_t) \tag{2}$$

where investment is proportionate to incremental output, and B_t is a matrix of incremental capital coefficients describing unit capital formation required per unit increase in output.

Equations 1 and 2 assume that the transformation of inputs into producing capacity is completed during a single time period —that is, the investment function has a uniform one-year lag. For most major plant and equipment expenditures, however, such an assumption is misleading. For example, it takes about three years to build a fossil-fueled generating station. A more complex lag scheme is required.

Investment in any one period will be the sum of components required to complete investment sequences begun in the past and initial investments of sequences contributing to increases in capacity years hence. This concept is illustrated by Figure 1.

The figure assumes for illustration that some particular piece of capital requires three periods to build. At the beginning of period t, the stock of available producing capacity is S_t. Over three periods, capacity increments of A (to S_{t+1} in period t),

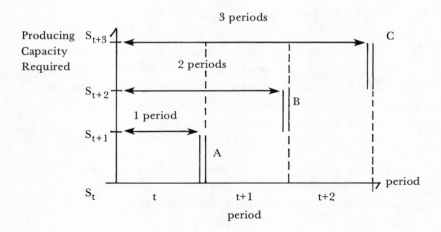

Figure 1. *Lag Structure*

B (to S_{t+2} in period t+1), and C (to S_{t+3} in period t+2) will be required. To produce increment C, some inputs to capital formation need to be produced during period t because of the assumed three-year formation period. Note that *not all* the inputs needed for C need to be produced during t, just those with three period lead times. Note also that *not all* of any particular input $[b_{ij}]$ in the three-year sequence must be produced during one period. For example, some fraction of the total input from the construction industry might be needed in each of the three periods of capital formation.

Similarly, inputs for increment B that have two-year lead times must be produced during t (those inputs for B with longer leads having already been produced during the previous period t−1). Any remaining input needs for increment A must also be produced during period t, in order to assure that a sufficient stock of capacity S_{t+1} will be available to meet next period's demands.

This conceptualization can be used to lag the incremental capital formation required to increase output over a number of periods back to some horizon N, the maximum construction period. Define a set of lag operator matrices $L_{0,1,\ldots N}$ such that

$$\sum_{n=0}^{N} \ell_{n,i,j} = 1, 0 \leqslant \ell_{n,i,j} \leqslant 1 \tag{3}$$

These lag operators specify the fraction of total capital inputs $(b_{i,j})$ that occur in each of the periods over the investment horizon. If some $\ell_{n,i,j}$ is zero, then no investment input from industry i to industry j occurs in the nth lag period. Similarly, if $\ell_{n,i,j}$ is one, then the total capital input to industry j from i is produced in period n. As noted above, values of ℓ between 0 and 1 are common. The $\ell_{n,i,j}$ are a unique function of each incremental capital coefficient $(b_{i,j})$ and are empirically observable. Although the lag operators are assumed time independent, such a restriction could easily be relaxed to take technological change into account.

The vector of total investment for any period t will be the sum of inputs undertaken during t for all future capacity increments:

$$K_t = \sum_{n=0}^{N} L_n \cdot B_{t+n} (X_{t+n+1} - X_{t+n}) \tag{4}$$

where $L_n \cdot B_{t+n}$ is the dot product.

Equations 1, 2, and 4 yield a general dynamic model with lagged investment:

$$X_t = A_t X_t + \sum_{n=0}^{N} L_n \cdot B_{t+n} (X_{t+n+1} - X_{t+n}) + C_t \tag{5}$$

This equation 5 is readily solvable by an iterative procedure if the time path of C_t is known and terminal conditions specifying the growth rate of output after the last period in the model are given. Assuming zero growth as the terminal condition, the general solution is given by:

$$X_t = G_t^{-1} C_t + \sum_{n=0}^{N-1} G_t^{-1} Z_{t,n} X_{t+n} + G_t^{-1} L_N \cdot B_{t+N} X_{t+N+1} \tag{6}$$

where
$$G_t = I - A_t + L_o B_t \tag{6a}$$
and
$$Z_{t,n} = L_n \cdot B_{t+n} - L_{n+1} \cdot B_{t+n+1} \tag{6b}$$

Equation 6 reduces to the Leontief dynamic inverse when there is a single period investment lag ($N=1$, $L_o = I$).

3.0 SPECIFICATION AND DATA

The most detailed U.S. input-output table available to implement the model developed in Section 2 is the 384-sector current-flows matrix for 1963, published by the Bureau of Economic Analysis (BEA). Available capital matrices, however, are at 83-sector levels of aggregation, corresponding to BEA current flow tables for 1958 and 1963. Even at 83 order, many sectors not of direct interest to capital formation studies are shown in detail. For example, agriculture is represented in sectors 1–4, and services in sectors 70–77. To facilitate computation and analysis, a 33-sector aggregation of the 83-order tables was developed. The aggregation was designed to retain maximum detail in capital-producing industries while eliminating detail in consumer-oriented industries. The correspondence between 33-order sectors and 83-order BEA tables is given in Table 1.

Table 1. *33-Sector Aggregation*

Sector	Name	BEA 83-Order Sector
1	Agriculture	1, 2, 3, 4
2	Metals mining	5, 6
3	Coal	7
4	Nonmetals mining	9, 10
5	Construction	11, 12

6	Food and tobacco	14, 15
7	Textiles	16, 17, 18, 19
8	Wood Products	20, 21
9	Furnishings	22, 23
10	Paper and printing	24, 25, 26
11	Chemicals	27, 28, 29, 30, 32
12	Petroleum	8, 31
13	Leathers	33, 34
14	Glass, stone, and clay	35, 36
15	Primary metals manufactures	37, 38
16	Structural metals	40
17	Other fabricated metals	39, 41, 42
18	Engines and turbines	43
19	Heavy moving machinery	44, 45
20	General industrial machinery	46, 49
21	Manufacturing machinery	47, 48, 50, 52
22	Industrial electrical apparatus	53
23	Electrical equipment	54, 55, 58
24	Electronics	51, 56, 57
25	Transportation equipment	13, 59, 60, 61
26	Miscellaneous manufacturing	62, 63, 64
27	Transportation and warehousing	65
28	Communications	66, 67
29	Utilities	68
30	Trade	69
31	Services	70, 71, 72, 73, 74, 75, 76, 77
32	Government	78, 79
33	Dummies	81, 82, 83
34	Fossil generation	—
35	Nuclear generation	—
36	Hydroelectric generation	—
37	Other generation	—
38	Transmission	—
39	Distribution	—
40	Administration	—

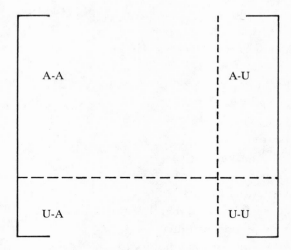

Figure 2. *Model Partitioned Matrices*

Modeling of the utility industry itself requires more detail than is available even from a 384-order table. To obtain additional detail, sector 68.01 (private electric utilities) and sectors 78.02 and 79.02 (public electric utilities) were expunged from the current and capital tables and then reincorporated as a set of seven subindustries, 34–40. The resultant 40-order input-output tables are partitioned between a 33-order description of the economy (partition A) and a 7-order description of the utilities (partition U) as shown in Figure 2. The seven utility subindustries correspond to fairly homogenous utility technologies:

34. Fossil-fueled steam generation.
35. Nuclear steam generation.
36. Hydroelectric generation, including pumped storage.
37. Other generation.
38. Transmission.
39. Distribution.
40. Administration.

Inputs from the rest of the economy to these processes are given in the A-U partition. Economic interchanges within the utility industry are given in the U-U partition. The U-A partition will be set to zero, as explained below.

The partitioned matrix is specified so that all electricity is produced in sectors 34–37, then "sold" to sector 38 (transmission), which in turn "sells" to sector 39 (distribution). The distribution subindustry, in turn, "sells" to the rest of the economy. By directly varying the input coefficients from generation to transmission and distribution, it is possible to represent any mix of the seven major utility technologies. This capability facilitates study of alternative utility investment strategies.

To reduce data requirements, a simplifying restriction was placed on the general model. It was assumed that all electricity is sold directly to final demand. Thus the partition U-A is set to zero and only a single projection of total consumption of electricity by the whole economy from sector 39 (distribution) is needed to derive the model. This simplification of the model also provides easily comprehensible results. When other final demands are set to zero, total outputs are induced solely by the interindustry requirements, direct and indirect, of electricity production. The time path of total output vectors shows absolute quantities of output required to support total electricity demand over time. Hence, impacts on the economy caused by utilities are distinguished from impacts caused by growth or change in other final demands.

Two principal sets of data were developed for this model. One set was for the 33-sector A-A partition. Current flows for 1963 came from the published BEA tables [1]. Incremental capital coefficients for 1958 and 1970–76 for the A-A partition were obtained from Battelle Memorial Institute [3]. Capital coefficients for 1963 were obtained from the Battelle data by linear interpolation. Since there are no 1980 tables, only 1963 current and capital coefficients were used for the results reported here. In effect, it is assumed that only technology *within* the utilities sector will change. Because of the general lack of information on lags, all investment for the A-A partition of the model is assumed to occur with single-period lags—that is, N=1 for this partition. It is recognized that the use of non-lagged 1963 data for sectors other than utilities limits the validity of conclusions about 1980.

A second set of data was developed for the utility partitions. Current and capital input coefficients for 1963 and 1980 for the A-U partition were developed by the author and reported in [4]. To simplify computation, only 1963 and 1980 coefficients were used. Linear interpolation of coefficients for other years would have been possible for the utility sector, but changes in coefficients between 1963 and 1980 were small. Given the lack of 1980 data for the rest of the economy, the additional refinement did not appear justified.

Lag operators for the utility partitions were also developed. The horizon was taken to be seven years (N=7). Except for construction inputs, the maximum lag for investment inputs to administration was taken as one year, and for transmission and distribution as two years. Some generating equipment items are ordered and installed within a year; for these items, $\ell_n = 1$ for some n. Inputs from other industries, for example construction and boiler fabrication, are phased over more than one year, usually with light initial spending and heavier inputs as the new plant nears completion. The timing and proportions of these inputs (the ℓ_n) were estimated from sources cited in [5], and are given in [6].

The final data requirement was a projection of total demand for electricity. The historical rate of growth in real output and in constant dollar revenues at 7.2 percent per year was used to extrapolate to terminal year 1987 from observed data. All data were expressed in terms of 1963 dollars.

4.0 RESULTS OF CHANGING TECHNOLOGY

This section uses the model described above to explore two

major alternative utility investment strategies. This is done by
selecting alternative investment strategies, implementing them
in the model, and comparing resultant alternative total output
paths. The two major alternatives examined are:

(1) Technical change within existing process mix;
(2) Technical change with changing process mix.

In other words, we are trying to discover what the economic
effects of improving existing processes are as compared to introducing new processes.

4.1 Changes within Existing Technologies

One strategy open to utilities that are confronted with exponen-
tially rising demand, increasing costs, and tightening environmental restrictions is simple improvement of existing technologies. Such improvements include (a) larger unit sizes, resulting
in economies of scale; (b) higher voltage transmission, resulting
in larger capacity per dollar of investment; and (c) pumped
storage plants that may reduce the cost of peak load energy.
Other modifications are being made in response to environmental criticism. More efficient fly-ash-collection equipment,
expanded use of cooling ponds or towers instead of once-through
cooling systems, and underground distribution systems are
examples. Technological changes affect utility capital and
operating inputs in diverse ways. To explore the interindustry
consequences of such changes, total output paths based on the
observed 1963 utility technological structures and generating
mix were compared with total output paths based on projected
1980 technological structures using the same mix of generating
technologies. The model was implemented for the period 1963
to 1987. The period of concern for the study was 1970–80, with
earlier years included for comparison of simulated versus actual
outputs. Given the maximum construction lag period of seven
years, continued growth through 1987 was assumed in order to

vitiate the effect of terminal conditions on the decade under
study.

Coal (sector 3) is a typical industry supplying current inputs
to utilities. Figure 3 shows the time path of coal output. There
is steady growth in total coal output through 1987, the terminal
year for the simulation. There is no decline in output, although
growth, and therefore capital formation, ceases after 1987. This
result indicates that direct demand for coal as a utility fuel
predominates over any indirect requirement in capital producing
industries. With either version of technology, the rate of growth
in coal output is 7.2 percent, as expected in a linear model. The
higher coal output in utilities with 1980 instead of 1963 technology (Figure 3) reflects three underlying differences in the
input structures. The first is the lower efficiency of units with
1980 technology due to higher turbine backpressures from artificial cooling systems. The second is an increased use of coal
relative to natural gas in fossil fuel plants. The third is higher
coal prices due to premiums on low sulfur coal. Premiums were
assumed to be about $1.50 per ton on 1963 prices (about $6/ton),
or 25 percent. This premium flows through directly to coal
industry gross revenues, increasing dollar output of coal.

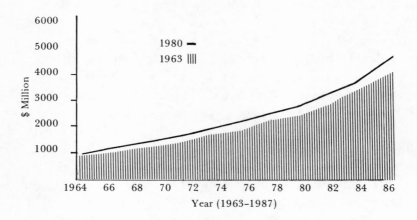

Figure 3. *Time Path of Coal Output (Sector 3)*

Industries supplying current inputs of utilities show steadily increasing outputs. Sectors displaying such output paths are those not concerned directly or indirectly with supplying capital needs of the utilities, and therefore not *directly* affected by alternative investment strategies. Investment decisions (as between coal and other fossil fuels or nuclear power), can, of course, affect subsequent current account purchases of these outputs. Industries supplying current inputs to utilities include sectors 3 (coal), 12 (petroleum), and 34–40 (electric power sectors). Sector 27 (transportation) displayed only a very slight decline to 1987.

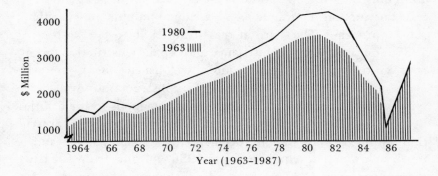

Figure 4. *Time Path of Construction Output (Sector 5)*

Intuitively, one would expect technological change to have noticeable effect on the construction industry (sector 5). Figure 4 illustrates this, showing that 1980 technology will require larger quantities of construction.* A large portion of this increase can be accounted for by the inclusion of artificial cooling methods, either ponds or towers, with costs ranging from $2–20/kwh of new capacity, depending on the system. Increased

---

*Construction industry input to utilities was more narrowly defined than for BEA tables. Design and prime contracting done by utilities are excluded from the sector 5 (construction) output, tending to reduce the industry's total output relative to BEA figures.

requirements from underground distribution also affect construction output.

Computed construction output from 1963 to 1970 nicely tracks observed purchases of construction by utilities, with about a one-year overall investment lag. This may indicate that indirect construction requirements in the rest of the economy (one-year lag), combined with one or two year lags in direct requirements for transmission and distribution, tend to predominate over the larger lags in generating plants.

Output projections past 1980 reflect terminal year conditions. The sharp decline in construction output from 1982 to 1986 illustrates the operation of the lag mechanism; less and less construction is required as the assumed terminal no-growth state approaches. The sharp increase from 1986 to 1987 is the 1987 requirement for maintenance and repair construction (given constant demand levels) which indicates "disinvestment" in new construction, specifically the freeing of construction output previously used indirectly by industries supplying goods to the utilities. Because the model does not distinguish between types of construction output, "disinvestment" occurs, reducing maintenance and repair below actual requirements in 1986. Outputs from 1981 to the terminal year 1987 are primarily of theoretical interest.

Terminal period behavior can be used to categorize industries according to their sensitivity to investment by utilities. When an industry supplies capital goods and/or current inputs to the utilities but nothing to industries that produce capital for the utilities, its output declines monotonically toward zero growth between 1981 and 1987. For industries selling primarily on current account, the decline is slight. In other industries the decline is large because most goods were sold on capital account. Examples of industries selling on current account include sectors 4 (nonmetals mining), 6 (food and tobacco), 20 (industrial machinery), 26 (miscellaneous manufacturing), and 28 (communications). Examples of industries selling on capital account are sectors 2 (metals mining), 17 (other fabricated metals), 22 (industrial electrical apparatus), and 23 (electrical equipment).

When the decline in output first dips below, and then rises to, the no-growth level, the sector is supplying capital goods to satisfy the direct requirement of the utilities and also the indirect requirements of other industries, indicating great investment impacts. Sectors in this category include 5 (construction), 9 (furnishings), 14 (glass, stone, and clay), 19 (heavy moving machinery), 21 (manufacturing machinery), and 25 (transportation equipment).

An illustration of how the model can be used to predict economic impacts is presented for structural metals (sector 16) in Figure 5. Although the output of the industry grows sharply

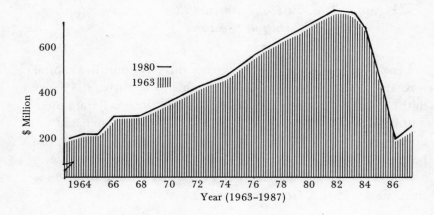

Figure 5. *Time Path of Structural Metals Output (Sector 16) with 1963 and 1980 Technologies*

with increasing utility demand, the sector is relatively unaffected by changes within utility technologies to 1980. Output declines as the terminal date approaches because most output from the industry goes directly to the capital needs of the utilities, some to indirect capital needs (the dip) and some to replacement (final level).

A final illustration, Figure 6, shows the impact of 1980 technologies on industrial electrical apparatus (sector 22). The decline to a 1987 output close to zero indicates that virtually

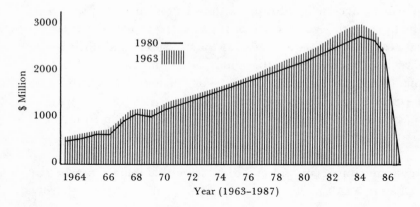

Figure 6. *Time Path of Industrial Electrical Apparatus Output (Sector 22)*

the entire output of this industry goes directly to utility capital formation. The industry is, therefore, highly sensitive to changes in the level of investment spending by the utilities. Total dollar output of sector 22 required to meet the growth of demand diminishes when 1980 input structures are introduced. This can be attributed to developments in EHV transmission, although increasing efficiency may be partly offset by a trend toward siting plants farther from load centers.

4.2 Results of Changing Process Mix

A second set of results bears directly on a key utility decision, whether to construct nuclear or fossil-fueled generating capacity. To a great extent this decision will be determined by cost of capital, construction period, fuel prices, and safety or other environmental considerations. It is unlikely that utilities acting individually will consider the impact of these investment decisions on the economy unless distortions or bottlenecks become large enough to affect costs and lags. Results of our computations indicate that bottlenecks may not be severe but there will be pervasive effects of importance to national policy makers.

Four different mixes of generating technologies were simulated for comparison. The percent of total generation for each technology is given in Table 2. Changes in the mix of generation

Table 2. *Generating Technology Mix*

Subindustry	Percentage of Total Generation			
Simulation	(1)	(2)	(3)	(4)
34 Fossil steam	80	75	65	60
35 Nuclear	5	10	15	20
36 Hydroelectric	10	10	15	15
37 Other generation	5	5	5	5
Total	100	100	100	100

technologies themselves are assumed to have no effect on outputs of transmission (38), distribution (39), and administration (40). The single most important finding of these comparisons is that changing the proportion of nuclear generation (hence investment) has major impacts on supplying sectors. Of the 33 nonutility sectors, only four—coal (3), petroleum (12), transportation equipment (25), and transportation (27)—have declining outputs as nuclear generation increases. All four industries are tied to the supply of fossil fuels whose consumption falls as nuclear generation increases. For some suppliers, the increase in output due to increasing nuclear capacity is fairly large; e.g., sector 8, lumber and wood products (construction materials); sector 11, chemicals (reactor moderator and fuel); sector 14, stone, glass, and clay (concrete); and sector 31, business services (insurance and legal). In most sectors, however, the increase is only a small fraction of total output.

Figure 7 depicts the change in the output of structural metals as the proportion of nuclear capacity increases. The changes are regular and gradual, perhaps less than might be expected given the metals requirement of reactor vessels. Increases in most

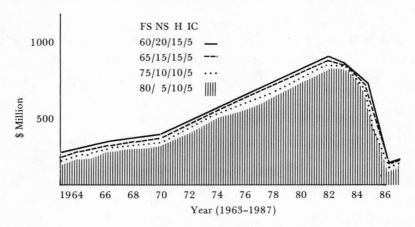

Figure 7. *Time Path of Structural Metals Output (Sector 16) with Different Mixes of Generating Technology*

other capital sectors, such as primary metals (15), industrial electrical products (22), even paper and printing (10), follow the same general pattern of gradual increases. It is interesting to observe that increasing nuclear capacity generates a greater demand for structural metals than do technological changes such as increased pollution controls or fossil-fueled units, the major alternatives to nuclear capacity.

Output of structural metals is not significantly affected by the increase in pumped storage peaking required with greater nuclear capacity. The increase in this hydroelectric generating capacity does significantly affect some other sectors, including construction (5). The effect is easily seen in engine turbines (18), shown in Figure 8. The major difference in this output comes not from increased nuclear capacity but from increased hydroelectric capacity. The cost of a pumped storage reversible turbine generator unit, used for peaking power, is higher per unit of effective output than that of a base-loaded steam unit.

One final inquiry into the uncertainties surrounding alternative investment strategies was made. To explore the impacts of varying investment lag structures, the maximum lag was shortened from seven to six to five years with suitably revised lag

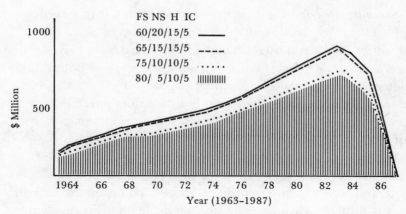

Figure 8. *Time Path of Engines and Turbines Output*
(Sector 18)

operators. Comparison of the output paths of sectors most affected by production of nuclear capacity (which had the longest lag) revealed only minor shifts in the time path of outputs. The most notable difference was the timing of the decline in output caused by the assumed terminal conditions. Before 1980 the assumed change in the lag structure had no significant impact.

5.0 CONCLUSIONS

It is possible to draw three conclusions from this study.

First, the general lagged dynamic input-output model is an effective tool for studying the interindustry effects of industrial growth when investment in plant and equipment is a major factor. Time paths of total outputs can be studied to determine major economic impacts. The principal limitations of such modeling is the difficulty of collecting data for detailed model specifications. As with any input-output analysis, the results are subject to all the limitations of linear production functions with fixed coefficients for each version of technology.

Second, although the principal changes within utility technologies involve significantly different demands on selected industries, in general these changes do not have a pervasive macroeconomic impact. For example, installation of pollution abatement equipment on fossil-fueled generating units affects sectors producing this equipment, but does not appreciably affect other sectors. Economy-wide investment requirements are not appreciably higher with the newer technologies.

Finally, increasing the proportion of nuclear capacity, a major alternative to fossil-fuel generating capacity, requires almost all sectors of the economy to produce more total output. Although the fossil-fuel/nuclear decision may be made on other bases, the macroeconomic implications should be considered by investment planners in government and in the utilities industry. Any bottlenecks that might arise as utilities' capital needs continue to expand will be even more probable and more severe with the nuclear alternative.

REFERENCES

1. U.S. Department of Commerce, Bureau of Economic Analysis, *Survey of Current Business* (Washington, D.C., Government Printing Office, November, 1969).
2. Federal Power Commission, *1970 National Power Survey* (Washington, D.C., Government Printing Office, 1972).
3. Fisher, W. Halder, and Cecil Chilton, *An Ex Ante Capital Matrix for the United States, 1970–1975* (Columbus, Ohio, Battelle Memorial Institute, March, 1971).
4. Istvan, Rudyard, *1980 Inputs for Private Electric Utilities* (Washington, D.C., Interagency Growth Project, August 1972).
5. —— *Generation: Technoeconomic Assessment of Electrical Generation Processes, 1963–1980,* Supplement to 4, above (Washington, D.C., Interagency Growth Project, August 1972).
6. —— "Outputs and Growth in the Electric Utilities," unpublished dissertation, Harvard University (1972).
7. Leontief, Wassily, "The Dynamic Inverse," in Anne Carter and Andrew Brody, eds., *Contributions to Input-Output Analysis* (Amsterdam, North-Holland Publishing Company, 1970).

4

Predicting Economic Impacts
of New Technologies:
Methodology and Application
to Coal Gasification

JAMES E. JUST

1.0 INTRODUCTION

The capital demands to feed our large and growing energy
appetite are enormous. Electric utilities alone account for one
sixth of new plant and equipment purchases. All energy sectors
account for over 10 percent of total construction. As natural gas
supplies dwindle, new technologies such as coal gasification or
imported Liquefied Natural Gas (LNG) will have to take over a
share of the market. The new energy technologies are extremely
capital intensive, especially when compared to what they replace.
At the same time these changes are taking place, our financial
institutions are faced with the capital demands by industry for
pollution-abatement equipment and by affluent young families
for housing. Can the United States satisfy such terrific capital
demand? Will there be a few or many bottlenecks? This depends
on how the actions of individuals and firms are linked to the
response of the economy as a whole.

The links between the labor, material, and capital demands of
the firm and the macroeconomic variables for GNP, total investment, and consumption are very complicated. Individual families,

government units, and purchasers of capital goods demand products and services in the market place. Suppliers respond by utilizing current technology and, in turn, make demands on other suppliers. The wages, profits, and taxes paid by these suppliers enable the consumers to pay for their goods.

Input-output analysis was developed to focus on these interrelationships. Its usefulness as a forecasting tool has been hampered by the birth of new industries, the introduction of new technologies, the changes in relative factor prices, and the difficulty of predicting consumer behavior. This paper describes a method of using engineering studies to help solve some of the problems associated with technology developments and improvements and new industries.

The technique utilizes engineering design studies undertaken for the construction and operation of the new technology or industry. Engineering information is converted into technological and capital coefficients for the new process. Next an assumption or exogenous calculation must be made regarding the rate of introduction of this new technology. This assumption allows the effects of the new process to be integrated with a general forecast of how the rest of the economy is growing. It helps the analyst to identify the differential impact of introducing or not introducing the new process.

As can readily be seen, the technique is not a panacea for all forecasting problems, but it does provide an encouraging new means of assessing the impacts of new technologies. The analysis can easily be augmented to include various pollution emissions, employment, or resource usages that are proportional to the total output of any particular industry, as will be illustrated later.

Our research utilizes a projection of the 1980 economy prepared by the Interagency Growth Project of the Bureau of Labor Statistics (BLS) [6]. These projections were incorporated into a model that contained environmental parameters and coefficients of new technology that had been derived from basic engineering studies. The research focused on the economic impacts of investing in these highly capital intensive new technologies and of day-to-day operation of such plants. Finally, a dynamic input-output model was used to make a series of 1985 projections. These projections involved different rates of energy use growth and were performed with and without the new technologies.

The major results document the sensitivity of total capital investment to changes in the growth rate of energy use and to the adoption of new technology. They also show that very small changes in the overall growth rate of personal consumption or government expenditures can restrain total investment to within its historical limits as a percentage of GNP. The significance of these results is that the people of the United States can sustain the huge investment demands created by rapid energy demand growth by reducing the growth rate of personal consumption and government spending by less than 0.1 percent per year through 1985. Overall GNP growth rate remains unchanged, because the sum of the growth rates of investment and noninvestment goods has been held constant.

2.0 THE GENERALIZED INPUT-OUTPUT MODEL

The generalized input-output model used in this study is illustrated in Figure 1. The core of the model contains the actual and projected input-output structures describing the 1963, 1970, and 1980 economies. This part of the model is discussed in more detail later.

The noneconomic quantities are referred to as accessory variables and are summarized in the bottom half of Figure 1. These are the outputs of the model. They are assumed to be proportional to the total output of each sector. For example, let S be the total emissions of SO_2 (or any other accessory variable) by the 1980 economy and let $E = [e_k]$ be the vector of coefficients describing the SO_2 emissions per dollar of total output for each sector. In other words, e_k is the SO_2 emitted per dollar

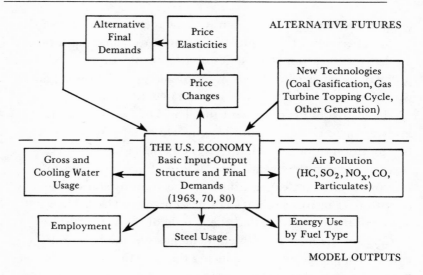

Figure 1. *Energy-Oriented Generalized Input-Output Model*

of output of the k-th industry. If X is the total output vector, then the total SO_2 emissions S is the inner product of X and E or

$$S = E^T X = X^T E. \tag{1}$$

Similar relationships hold for the other accessory variables.

The boxes in the upper half of Figure 1 represent the various means of interacting with the model. These boxes represent the alternative future being investigated. A scenario for the future can include changes in technology and in size and composition of GNP.

Alternative futures are specified by developing a final demand vector to represent the conditions of the scenario and modifying the technological and capital coefficients to include the amount and kind of new technology that is specified. Once these changes are made, the total outputs and investment requirements can be calculated. The values of the accessory variables are then obtained by simple multiplication as indicated above.

When making projections of final demands, it is necessary to calculate the investment required to support that level of final demand. Since final demand itself depends on the investment level, it is necessary to make sure that assumed levels of investment and other final demands are consistent. A simple two-period model can be used to illustrate the iterative technique used here. Assume that (1) the same technological coefficient matrix A applies for both periods; (2) total final demand Y consists of final demand purchases by households and governments (Y^F) and capital investment purchases by all sectors of the economy (Y^I), or

$$Y = Y^F + Y^I \tag{2}$$

and (3) the capital matrix C is defined as $C = [c_{ij}]$, where c_{ij} is the marginal capital purchase from sector i by sector j required to expand the capacity of sector j by one dollar of output. Thus, if X_o were the total output in period T_o and X_1, the total output in period T_1, the total new investment required is $C(X_1-X_o)$.*

The objective is to find for period t_1 the total output (X_1) and total final demand (Y_1), given X_o, the total output in period t_o, and Y_1^F, the noninvestment final demand in period t_1. The model assumes that sectors always operate at 100 percent capacity, so that output can only be increased by capital investment.†
The basic equations for this model are:

$$X_1 = (I-A)^{-1} Y_1 = (I-A)^{-1} (Y_1^F + Y_1^I), \tag{3}$$

and

$$Y_1^I = + C(X_1 - X_o). \tag{4}$$

These equations can be solved for total output (X_1) and total final demand (Y_1):

---

*The assumption that investment is proportional to the change in output is reasonable only for positive changes in output. Since all growth rates are positive [i.e., all $x_{1i} > x_{oi}$] in this study problems of negative investment do not arise.

†Slack variables can be used to modify this assumption, but were not in this study.

$$X_1 = (I-A-C)^{-1}(Y_1^F - C X_o); \tag{5}$$

$$Y_1 = Y_1^F + C(X_1-X_o). \tag{6}$$

Equations (5) and (6) can be used to investigate the effects on investment Y^I and total output X of changes in the growth rates of individual components of Y^F.

Alternative projections of 1985 were made, all based on the same total GNP.* The composition of GNP varied with the projected composition of future technologies. Figure 2 describes the projection model used in this study to ensure a constant

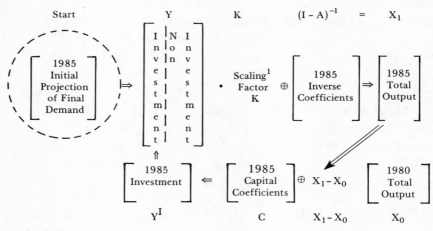

Symbols
- · Scalar Multiplication
- ⊕ Matrix Multiplication
- ⇒ Equals

[1] Scaling factor is chosen so that GNP = $|Y| = \Sigma_i Y_i = \$1.34$ trillion.

Figure 2. *Quasi-Dynamic Projection Model*

$$*GNP = |Y| = |Y^F + Y^I| = \sum_{i=1}^{N} Y_i \text{ where N = number of sectors in model}$$

and the magnitude signs indicate arithmetic additions of the vector elements.

1985 GNP of $1.34 trillion† (1958 dollars). It is a slight modification of the above two-period analysis. Convergence of the iteration can be assured by modifying the scaling factor.

3.0 NEW TECHNOLOGIES

3.1 Derivation of Technological Coefficients

Our example for deriving technological coefficients for new technologies is taken from a report by the Institute of Gas Technology (IGT) [4]. This report describes a 500 billion BTU/day gasification process that operates via hydrogasification and electrothermal gasification of lignite. This example was used because of its ready availability, not because it is necessarily the most likely future gasification process.

The IGT report summarizes a preliminary engineering study undertaken to provide estimates of the capital and operating costs of the plant and thus to determine the price at which the output gas could be sold. Such studies can be made at various levels of detail depending upon whether the purpose is to develop preliminary estimates of output prices or to design and construct the actual plant. The conceptual procedure is the same in both cases, but the level of detail differs substantially.

This basic procedure consists of

(1) identifying the major individual (unit) processes that make up the overall flow of materials through the plant (e.g., coal receiving, storing, and preparation)

---

†This GNP represents a 4.4 percent per year growth rate from the BLS projection of the 1980 GNP. It was calculated by excluding any contribution from Bureau of Economic Analysis sectors 84, 85, and 86 (Government Industry, Rest of the World Industry, and the Household Industry respectively). These dummy sectors were excluded because they do not interact with other sectors; they only contribute to GNP.

(2) determining the size of the desired plant in terms of daily output (this is an executive decision)

(3) estimating the cost of equipment for each process to achieve the required daily output

(4) estimating the construction time (which determines interest charges during construction)

(5) estimating the cost of integrating and connecting the various processes

(6) determining the total construction cost (the sum of all of the above plus land costs, contractor's profit and contingency allowances).

Once the total construction costs are known, the unit price of the output is calculated by

(1) estimating the number (and cost) of men required to operate each processing section of the required size

(2) estimating maintenance and supply costs (usually as a percentage of total construction costs)

(3) calculating raw material costs for the desired yearly output

(4) estimating overhead, taxes and depreciation charges and profits (percentage of the above figures)

(5) summing these figures to get a total yearly cash requirement and finally

(6) dividing this total by the total yearly plant output in BTU's to obtain a unit price in dollars per BTU.

The knowledge of what types of process equipment (e.g., carbon versus stainless steel) are required for a particular plant is usually derived by examining similar plants. For a new process there are obviously no similar plants, and this introduces a potentially large source of error. Engineering judgment must be used in these studies. As a project of this type advances from the conceptual to the laboratory to the pilot plant stage, the engineering cost studies become much more exact. By the time a demonstration plant is completed, almost all of the uncertainty is gone.

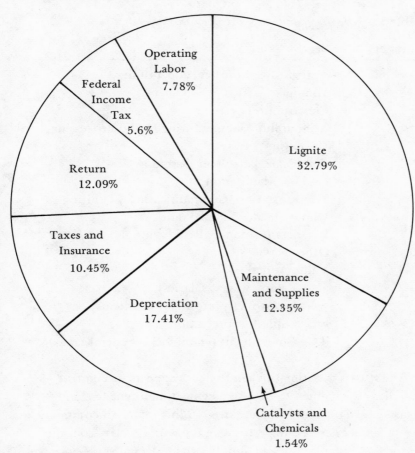

Source: Tsaros [4], p. 67.

Figure 3. *Components of Pipeline Gas Price*

The outputs of these cost studies are detailed lists of equipment, material, and labor requirements. These lists can be summarized as pie charts illustrating the various elements that make up the price of the output. Figure 3 does this for the price of synthetic natural gas as described in the IGT report. This chart will be used to describe the derivation of new technological coefficients. Table 1 lists eleven sectors used in the example. The actual study employed a 110-sector model.

Table 1. *Hypothetical Ten-Sector Economy*

Number	Sector Name
1	Agriculture, Forestry, and Fishing
2	Mining
3	Construction
4	Nondurable Manufacturing (Food Processing, Textiles, etc.)
5	Chemicals, Petroleum Refining
6	Durable Manufacturing
7	Transportation, Communications, Utilities
8	Wholesale and Retail Trade
9	Finance, Insurance, Real Estate
10	Other Services
11	Value Added
	a. Labor (wages, salaries)
	b. Investors (interest and dividends)
	c. Capital Depreciation
	d. Government (state, local, Federal taxes)

A first pass at estimating coefficients appears in Figure 4. Supplies are assumed to be 15 percent of maintenance, and insurance 10 percent of local taxes. In this figure all commodities or services are assumed to be purchased directly from the sector that manufactures or supplies them. Retail trade and transportation are ignored in this round. For example, catalysts and chemicals are assumed to be purchased directly from the chemical manufacturing sector even though they may have been purchased from a local distributor.

The convention followed in input-output analysis is that wholesale and retail trade do not purchase any goods for resale. Instead, the purchaser is shown as having bought any particular good directly from the manufacturer at the producer's price (i.e., what the manufacturer receives from a wholesale buyer) *and* paying the trade margin or markup directly to the wholesale and retail trade sector. Thus any transaction is recorded as

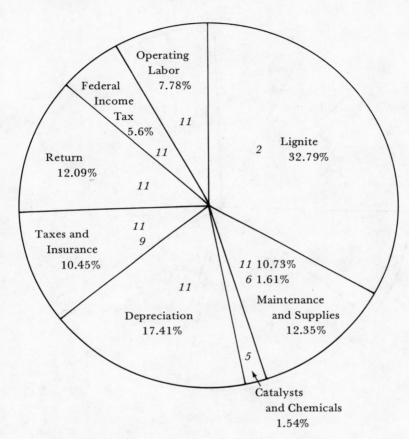

Figure 4. *Components of Pipeline Gas Price*
Preliminary Assignment of Sectors

two separate entries, one to the manufacturing sector and one to the trade sector.

Transportation charges are handled similarly to trade margins. The purchaser is shown as paying the transportation charges directly to the transportation sector. Figure 5 applies these concepts to the IGT example. Here 25 percent of the price of lignite is assumed to be transportation charges. No trade margin for lignite purchases is included because the company is assumed to buy directly from the mine. Supplies and catalysts and chemicals

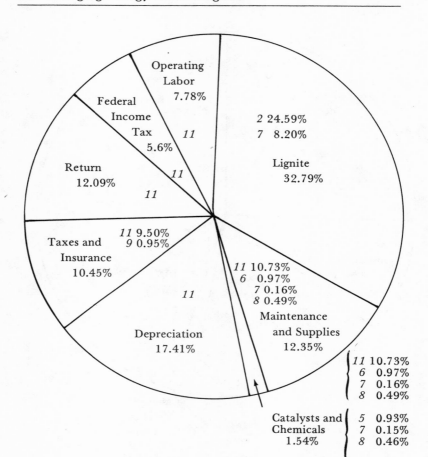

Figure 5. *Components of Pipeline Gas Price Final Assignment of Sectors (Including Trade and Transportation Margins)*

are assumed to have a 30 percent trade margin and a 10 percent transportation margin.

All that remains is to collect and sum items assigned to each category.

What we have referred to as technological coefficients in this paper and in the input-output literature might more properly be called operating input coefficients. "Technological coefficients" is a misnomer, since these coefficients describe the technology of the industry only in a crude and implicit sense. There certainly is no danger of revealing trade secrets from this approach. The operating input coefficients are analogous to the ingredients list in a cooking recipe. By combining all of these inputs in some artful way, a car, transistor, etc., results.

For new technologies such as coal gasification, there are many competing processes that perform the same function or yield the same product. In developing the technological coefficients to represent such technology, it is important to make sure that the coefficients are representative of all the processes that will be used and to find out whether any conclusions from such a study are sensitive to the exact mix of processes chosen.

3.2 Derivation of Capital Coefficients

The process involved in deriving capital coefficients is similar to that for the technological coefficients. The basic strategy is to assign each piece of equipment to its producing industry; remove the transportation and trade margins; allocate construction, insurance, engineering, interest, etc., charges; and divide the total purchases from each section by the total cost of the plant. The procedure yields a vector the elements of which sum to 1.0 and which can be used to allocate each dollar spent on coal gasification plants to the industries where the capital goods originate.

Given the proportions of capital coming from each industry of origin, the capital coefficient vector can be found by multiplying the vector of proportions by the capital/output ratio that describes the dollars of capital investment in plant per dollar of product output from the plant. This is calculated easily by dividing the capital cost of a new plant by the value of its yearly output.

These are the procedures that were followed to derive capital and technological coefficients for all the various new energy technologies that are discussed later. Detailed derivations and assumptions can be found in my paper, "Impacts of New

Energy Technology Using Generalized Input-Output Analysis"
[2].

3.3 Incorporation of New Technology

The new technologies were incorporated into the input-output
framework using the following scheme. Suppose the old technological process for sector i (e.g., natural gas production) is
represented by the technical coefficients vector A_i^o and capital
coefficient vector C_i^o.* Next let the new technological process
(e.g., high BTU coal gasification) be A_i^N and C_i^N. If the new
technology is expected to take over a fraction g of the total
production of sector i and a fraction h of total capital investment by sector i, then the new technical coefficients are

$$A_i = (1-g)A_i^o + gA_i^N \tag{7}$$

where g = fraction of total production supplied by new technology and the new capital coefficients are:

$$C_i = (1-h)C_i^o + hC^N \tag{8}$$

where h = fraction of additional capacity made up of new technology.

These coefficient column vectors then replace the old ones in
the technical and capital coefficient matrices.

4.0 1985 PROJECTIONS

The results were obtained by projecting a series of five alternative
1985 futures involving various energy use growth rates, both with

---

*Thus the whole technological coefficient matrix could be represented as
the partitioned matrix $A = A_1 : A_2 : \ldots : A_n$. A similar partition holds for
the capital coefficient matrix.

and without new technologies. These will be referred to as the Low, Medium, High, High plus Hygas, and High plus Hygas plus Gas Turbine futures, and are defined below.

All of the projections used the 1980 technical coefficient matrix [6] with some modifications of the energy sectors. The investment component of final demand was recalculated for each projection using the 1975 Battelle capital matrix [7] modified slightly for the new energy technologies. The initial final demand projection for each alternative differed *only* in the amount of oil, natural gas, and electricity purchases.

The medium energy use growth rate future assumes a continuation of the 1970–80 final demand growth patterns and no change in industrial technology from 1980. The high energy use future reflects a 4 percent higher final demand (than the medium future) for oil, natural gas, and electricity and increased industrial consumption of electricity, gas, rubber, and plastics (reflected in slight increases in the energy rows of the technical coefficient matrix). These changes assume increased air conditioning and electric heat, worse gasoline mileage, and longer yearly driving distances. All of these projections assume that there will *not* be a supply limitation on natural gas and that the same domestic to foreign crude and natural gas ratios apply in 1985 that were projected for 1980.

The low energy use future involved 6 percent lower final demand (than the medium future) for oil, natural gas, and electricity and better conversion efficiency for electricity conversion and transportation. Two alternative high energy growth futures involving new technologies were also investigated. These technologies are described in Table 2. The High plus Hygas future included the introduction of high-BTU coal gasification (Hygas), while the High plus Gas Turbine future included Hygas and the gas turbine topping cycle (supported by low-BTU coal gasification). These technology modifications are described in Table 3.

The projection procedure aimed at a GNP of $1.34 trillion*

*This GNP represents a 4.4 percent growth rate from the projected 1980 GNP.

Table 2. *New Technologies Investigated*

High-BTU Coal Gasification (1000 BTU/SCF)

Process:	Electrothermal Hydrogasification (Hygas)
Data Source:	*Electrothermal Hygas Process-Escalated Costs* [4, 5]
Originator:	Institute of Gas Technology
Efficiency:	71.7%
Nominal Plant Size:	500 million SCF/day (90% load factor)
Nominal Cost:	Plant—$310–354 million[a]
	Gas—54.8–72.4c/10^6 BTU

Low-BTU Coal Gasification (173 BTU/SCF)

Process:	1980 Texaco Partial Oxidation (Hot Carbonate Scrubbing)
Data Source:	*Technological and Economic Feasibility of Advanced Power Cycles* [3]
Originator:	United Aircraft
Efficiency:	87%
Nominal Plant Size:	842 million SCF/day (70% load factor)
Nominal Plant Cost:	Plant—$27.5 million[b]
	Gas—17.6c/10^6 BTU

Gas Turbine Topping Cycle (Combined Gas and Steam Cycle or COGAS)

Process:	1980 High Inlet Temperature (2800°F) Turbine with Waste Heat Boiler Steam Cycle (Using Low BTU Gas)
Data Source:	*Technological and Economic Feasibility of Advanced Power Cycles* [3]
Originator:	United Aircraft
Efficiency:	54.5%[c]
Nominal Plant Size:	1000MW (70% load factor)
Nominal Cost:	Plant—$94 million
	Electricity—5.3 mills/kwhr

[a]All dollar figures are in 1970 dollars.
[b]Includes working capital.
[c]Only the efficiency of the COGAS cycle. Overall efficiency is obtained by multiplying the two efficiencies.

(1958 dollars) in 1985 for all five alternative futures. This was accomplished with a balancing procedure using the model of Figure 2.

Table 3. *New Technology Modifications*

	Capital	Operating
Hygas (coal gasification)[a]	25% of new capacity (gas) additions will be in form of coal gasification.	5% of natural gas demand supplied by coal gasification.
Gas turbine topping cycle (combined with Low-BTU coal gasification)[b]	50% of fossil generation (15% of total generation) capacity additions will be added in form of gas turbine topping cycle.	38% of fossil generation (23% of total generation) will be by gas turbine topping cycle.

[a] High + Hygas Future: High Future is modified by the above addition of high-BTU coal gasification (the IGT Hygas process).
[b] High + Hygas + Gas Turbine Future: High Future is modified by the addition of both new technologies indicated above. Note that low-BTU coal gasification is used in conjunction with the gas turbine.

The major assumption in this procedure was that all sectors had the same income elasticity so that a constant scaling factor could be applied to all final demand purchases. This is a bad assumption for such industries as food and kindred products, but since the conclusions of this study are based on a differential analysis of the various projections and not on the absolute numbers involved, this assumption is not a major problem.

The salient points of the various balanced 1985 projections are summarized in Table 4. Total investment becomes a larger percentage of the 1985 GNP as energy use increases from low to high. The introduction of high-BTU coal gasification further increases investment, while the introduction of the gas turbine topping cycle (with or without low-BTU coal gasification) decreases it. The output of coal mining is seen to increase dramatically with the introduction of coal gasification. The three capital producing industries cited for illustration (plumbing, etc.)

Table 4. *Balanced 1985 Projections (1958 Dollars)*

	Low	Medium	High	High Plus Hygas	High + Hygas + Gas turbine
GNP					
(Billions of dollars)	$1340.8	$1343.0	$1339.0	$1340.9	$1341.0
PCE (% of GNP)	70.2	70.0	69.6	69.3	69.4
Investment (%)	16.6	16.8	17.5	17.7	17.5
Government (%)	13.8	13.8	13.5	13.6	13.6
Total output					
(Billions of dollars)					
Coal Mining	5.0	5.1	5.2	6.5	6.6
Plumbing, Structural Metals	18.2	18.5	19.3	20.0	19.7
Engines & Turbines	7.5	7.6	7.9	8.0	8.0
Construction Equipment	11.1	11.5	12.5	12.9	12.6
Private employment					
(Millions)	99.2	99.2	99.2	99.2	99.2
Air pollution					
(Million tons)					
Particulates	48.6	49.0	50.0	50.2	50.1
Hydrocarbons	91.7	92.2	92.3	92.3	92.1
SO_2	75.2	76.1	78.2	78.2	78.2
CO	122.7	123.9	124.8	124.8	124.2
NO	30.4	31.8	32.6	32.6	32.5
Steel use					
(Million tons)	194.0	195.0	198.1	199.6	198.6
Water use					
(Trillion Gallons)					
Gross	278.1	280.6	286.7	291.2	266.5
Cooling	126.0	128.3	134.3	137.8	117.8
Energy use					
(10^{15} BTU					
Coal	24.9	25.3	26.0	28.5	28.5
Oil	43.0	43.9	44.5	44.4	44.4
Gas	46.1	46.7	48.5	48.5	48.2
Electricity	33.0	33.8	34.9	34.8	34.8

respond to different energy use growth rates more than total investment as a whole. Total employment is approximately constant, but there is no indication of how the required skills might change. Certainly more people will be employed in construction and in the capital goods industries for the higher energy growth rate scenarios. Air pollution and steel usage behave as expected. The large decrease in water usage caused by the introduction of the gas turbine topping cycle results from the fact that the gas turbines are air cooled and that the conversion efficiency is higher than the standard generation plant.

The most important fact concerning these balanced projections is not found in Table 4. The noninvestment components of the balanced final demand projections were within 0.3% of the initial projections. In other words, only a very slight change in personal consumption and government expenditures was enough to balance the investment demands of the rapidly growing energy sectors. It seems unlikely that most sectors would notice a difference in sales of 0.3% over a five-year period. If most sectors were growing at 4.4% per year (the rate at which GNP is projected to grow) a decrease of 0.3% in sales would decrease the growth rate to 4.35%, hardly a significant change.

5.0 CONCLUSIONS AND FURTHER RESEARCH

(1) Total investment in general and capital good industries in particular (primarily turbogenerator manufacturers, boiler makers, and construction equipment manufacturers) are quite sensitive to energy use growth rates (especially electricity).

(2) The major impacts of introducing the new energy technologies will be on the capital goods industries listed above. Operation of the new plants significantly affects only coal mining.

(3) Introduction of high-BTU coal gasification will aggravate the demand for investment funds, and introduction of the

second generation gas turbine topping cycle (with or without low-BTU coal gasification) will decrease that demand.

(4) Slight changes in the overall growth rates of total personal consumption expenditures and government spending result in large fluctuations in total investment.

(5) If high energy growth continues and if investment is to remain within its historical limits as a percentage of GNP, energy investment will become a larger and larger part of total investment.

(6) Although interest rates are assumed to balance the supply of and demand for investment funds, the very act of saving more money (which is induced by higher interest rates) means that less can be spent on consumption goods. This in turn lessens the demand for investment funds because the growth rates of consumption sectors are lower. This indirect effect of interest rates on investment may be quite important.

The policy implications of these types of results can be important. Different sectors of the economy respond differently to changes in the interest rate. Housing construction seems to be particularly sensitive to interest rates. Knowledge in advance of what investment demands are likely to be provides additional information for planning government spending and taxes. Certainly more work on consumer and industrial response to interest rate changes needs to be performed. Will enough skilled construction labor be available to build all of the new required energy facilities? Manpower training programs can be developed if the need for such labor can be predicted long enough in advance. The generalized input-output model is, in fact, applicable to all of the above questions, either in pointing out the need for policy or in analyzing the effects of new policy. While the major government policy variable represented in the generalized input-output framework is government spending (broken down by sectors), the outputs provide insights into the possible effects of other types of policy decisions like manpower training.

It is the ability to incorporate engineering studies into the generalized input-output framework that negates many previous objections to input-output analysis. Engineering studies can be used to determine how technology is likely to change if relative price changes or if some fuel becomes unavailable or how technology may improve with time. More work is needed to improve technology forecasting, but the potential payoff is high.

REFERENCES

1. "Energy Crisis: Are We Running Out?" *Time* (June 12, 1972), pp. 49–55.
2. Just, J. E., *Impacts of New Energy Technology Using Generalized Input-Output Analysis*, Energy Planning and Analysis Group, M.I.T., Cambridge, Report 73-1 (January 1973).
3. Robson, F. L., et al., *Technological and Economic Feasibility of Advanced Power Cycles and Methods of Producing Non-Polluting Fuels for Utility Power Stations*, United Aircraft Research Laboratories (December 1970).
4. Tsaros, C. L., et al., *Cost Estimate of a 500 Billion BTU/Day Pipeline Gas Plant via Hydrogasification and Electrothermal Gasification of Lignite*, Institute of Gas Technology for Office of Coal Research (August 1968).
5. Tsaros, C. L., and T. K. Subramanian, *Electrothermal Hygas Process, Escalated Costs*, Institute of Gas Technology for Office of Coal Research (February 1971).
6. U.S. Department of Labor, Bureau of Labor Statistics, *Patterns of U.S. Economic Growth*, Government Printing Office, BLS Bulletin 1672 (1970).
7. Battelle Memorial Institute, *An Ex Ante Capital Matrix, 1970–75*, Battelle Memorial Institute (Columbus, Ohio, 1971).

Pollution & Abatement

Problems

5

Economic and Environmental Implications of Paper Production and Recycling

FRANS KOK

1.0 INTRODUCTION

Until the late 1960's paper recycling was a noncontroversial activity and a very stable industry. From 1947 to 1963 the total quantity of wastepaper recycled increased from 8 million to 9.6 million tons. During the same period the output of the paper industry nearly doubled, climbing from 21 million to 39 million tons. Clearly, the recycling industry was not in the mainstream of the expanding economy.

By 1965 the disposal of solid waste was recognized as a national problem. In that year Congress passed the Solid Waste Act and in 1970 expanded its provisions with the Resource Recovery Act. Later in 1970 President Nixon directed the General Services Administration (GSA) to revise its paper specifications. Increased recycle content would be required wherever feasible. The GSA now requires a minimum recycle content of 3 to 50 percent, with the exact percentage determined by the nature of the product. To divert the thrust of the new requirements, however, the paper industry has interpreted "recycle content" to include wood mill wastes, such as chips and sawdust. These were already being used in vast quantities by the industry prior to the 1970 revision of GSA standards.

In the late 1960's a special National Academy of Sciences committee working with the United States Bureau of Solid Waste Management recommended that recycled materials account for 50 percent of the projected national growth in paper and paperboard output through 1985. The implementation of this recommendation would result in a 35 percent recycling rate by 1985. Approximately 12 million tons of wastepaper were recycled in 1970, a rate of 22 percent. Since the 1985 paper and paperboard output is expected to double the 1970 level, 35 million tons of wastepaper would have to be recycled in that year to raise the recycling rate to the level recommended by the Academy. This would mean an annual growth rate of 7 percent for the recycling industry—not an exceptional rate for an ordinary activity in the economy but sizable compared to the industry's past performance. In the early 1970's an expansion of paper recycling by about one million tons per year would be required. By the early 1980's the rate of expansion would have to reach 2–2.5 million tons per year.

How would the economy and the national environment be affected if the paper industry were to rely increasingly on wastepaper rather than forests for its supply of fibers for the manufacture of paper and paperboard? The present study is an attempt to answer this hypothetical question.

At the present time there is no sign that the paper industry is planning to expand its recycling capacity to attain the goals recommended by the National Academy of Sciences. The American Paper Institute has estimated that the rate of growth in wastepaper consumption in paper and paperboard manufacture will be 3 percent in 1973 [6], just sufficient to maintain the present rate of recycling.

The pulp and paper industry may, however, be forced to respond more energetically to a different problem: industrial pollution. In 1964 the industry was responsible for 23.5 percent

of the industrial wastewater effluent discharge in the Northeast.
In the Southwest the industry figure was 26.4 percent; in the
Pacific Northwest, 16.7 percent [2]. Any serious effort to clean
up the environment obviously must include pulp and paper
manufacturers. A large-scale pollution abatement program in
the industry will also affect other United States industries and
resources. Prices of pulp and paper products will rise. Industries
supplying abatement inputs will be called upon to produce more.
In this study the economic impact of increased pollution-abatement activity within the paper and allied products industry will
be quantified.

The purpose of this investigation is therefore twofold: to
assess the impact of (1) increased paper recycling and (2) intensi-
fied pollution-abatement activities by the paper and allied products industry. The data analyzed were obtained from an input-
output model. The general equilibrium model used is described
in Leontief's "Environmental Repercussions and the Economic
Structure: An Input-Output Approach" [47]. Part 2 provides a
brief discussion of the present extent of pollution and pollution
abatement in the industry. In Part 3 auxiliary data on pulp paper
and recycling sectors are developed and aligned with the 83-order
input-output matrix prepared by the Office of Business Economics (OBE). In Part 4 the computed effects of an increase in
recycling are analyzed. Part 5 quantifies the effect of higher levels
of pollution abatement on the total economy and the environment.

2.0 POLLUTION AND POLLUTION ABATEMENT
IN THE PAPER INDUSTRY

2.1 Pulping Processes

The major sources of air and water pollution in the paper and
allied products industry are the various pulping processes. Once

the pulp enters the paper- and paperboard-making facilities, relatively few additional pollutants are emitted. Pulping is the process by which the cellulose fibers of plants are separated from each other. Many fibers are suitable for pulping and paper-making (e.g., straw, bagasse, esparto, bamboo), but in the United States the pulp industry has traditionally used wood to manufacture its products. Eighty-nine percent of the virgin pulp required in this country is manufactured by four basic processes; the remainder is produced by other processes or im-ported from Canada and the Scandinavian countries. The four major processes are:

Sulfate (Kraft) pulping	58.2%
Groundwood pulping	12.1%
Sulfite pulping	9.7%
Neutral Sulfite Semichemical pulping (NSSC)	8.9%

The percentages represent the relative tonnage of each process in the early 1960's. Since then the percentage for sulfate pulping has grown even larger, while percentages for sulfite and NSSC pulping have declined.

The groundwood pulping process is entirely mechanical and does not normally require any chemical inputs. Barked round wood or steam softened chips are reduced to fibers by grinders under a flow of water. The process breaks and ruptures the fibers without removing the nonfibrous material (lignin). As a result the paper produced from groundwood pulp discolors rapid-ly and is brittle. From an environmentalist's point of view, how-ever, groundwood pulping is the most desirable process: it pro-duces virtually no air pollution and less water pollution than sulfate, sulfite, or NSSC pulping, in which three processes the wood is cooked in a chemical solution. Disposal of this waste liquor is the major source of water pollution. In the sulfate process the heat value of the spent liquor is recovered by burn-ing. As a result sulfate pulp mills are very heavy air pollutors.

When wastepaper is recycled, it is treated by a chemical

process that de-inks it. Disposal of the cooking liquor, like disposal of the spent liquor from the virgin fiber chemical-pulping processes, poses a serious pollution problem. Later we shall examine certain pollution trade-offs between pollutants emitted by chemical virgin-fiber pulping processes and de-inking processes used in wastepaper recycling. When an ink-dispersion process is used to repulp wastepaper, the pollution problem is reduced significantly. But the dispersion process produces a greyish pulp, the familiar inside of cereal boxes. It has few other uses. After wastepaper has gone through the repulping process, it is used in the manufacture of paper, paperboard, or construction board in the same way as virgin fiber pulp. The pollution problem is relatively insignificant after the repulping process.

2.2 Measurement and Abatement of Pollutants

2.2.1 Water Pollution

Water pollution is measured in pounds of five-day biochemical oxygen demand (BOD_5) and pounds of suspended solids. BOD_5 is a laboratory measure for the quantity of oxygen, in pounds, that a newly released organic substance demands from its environment during its first five days. If the oxygen is plentiful in the environment, as it should be in an unpolluted stream, most organic substances will oxidize to a stable state within five days. Instead of the BOD_5 measure, some sources use population equivalent, the five-day oxygen demand of the waste discharged daily by one person. This has been established at 0.167 pounds BOD per day. Air pollution is measured in pounds of ambient particles and pounds of gases emitted.

Abatement of BOD_5 and suspended solids pollution is a matter of letting nature do its work in large lagoons before the effluent is discharged into a public body of water. The suspended solids settle out in these ponds and the BOD_5 is removed by bacteriological activity. The removal of suspended solids can be speeded up by coagulation. Bacteriological activity can be increased by aeration, introduction of cultures (activated sludge), addition of nutrients, and the maintenance of a proper pH range. Waste-water treatment is divided into four processes: pretreatment; primary treatment; secondary treatment; and tertiary treatment. In pretreatment, the waste effluent is prepared for the primary settling pond. Any usable fibers present in the effluent are caught in "savealls" and a proper pH is assured by adding alkalis or acids. Primary treatment of effluents removes up to 95 percent of the suspended solids. A certain amount of BOD_5 is also removed in the process. The settled solids are disposed of in landfills or by sludge burning. Secondary treatment also removes considerable amounts of suspended solids and is more effective for removal of BOD_5. Most effective is the activated sludge process, with a maximum BOD_5 removal of 95 percent. The aerated lagoon also performs well, with 85 percent reduction of BOD_5. Tertiary treatment is a catch-all term for treatment other than pretreatment and primary or secondary treatment. Color removal falls into this category. The technology of tertiary treatment is very new, and few mills engage in it at all.

2.2.2 Air Pollution

Ambient particles are removed from the air by electrostatic precipitators or fabric filters, both highly efficient processes. The latter remove up to 99.9 percent of all ambient particles. The removal of harmful and odious gases is more difficult. Various types of scrubbers are used but few do a good job. Theoretically, up to 90 percent of the hydrogen sulphide can be captured by Venturi scrubbers. In actual operation most scrubbers are less than 30 percent effective. They capture none of the special sulfur compounds (mercaptans) and few of the sulphur oxides. As a result, substantial quantities of gases are emitted by sulfate pulping mills. No sulfur abatement technique has yet proven to be both effective and applicable on a wide scale.

3.0 DEVELOPMENT OF DATA

3.1 Input Coefficient of the Paper Industries

On the detailed level of classification, the OBE has assembled input structures of seven pulp, paper, and paper conversion industries:

(24.01) Pulp mills
(24.02) Paper mills, except building paper
(24.03) Paperboard mills
(24.04) Envelopes
(24.05) Sanitary paper products
(24.06) Wallpaper and building paper and board mills
(24.07) Converted paper products not elsewhere classified (n.e.c.), except containers and boxes.

Input-output statistics are collected on an establishment basis in accordance with the principal-product rule. For example, a plant that produces an output mix that includes 50 percent or more paperboard will have *all* of its inputs and outputs assigned to sector 24.03, although it may actually produce a variety of products. The inputs used to produce products other than paperboard for this plant will distort the estimated input structure of paperboard production. When significant changes in the mix of products within a particular industry classification are analyzed, or when industry and technical information are to be combined, it is important to "purify" the technological input coefficients of that industry, i.e. to estimate what the input structure would be for the principal product alone rather than for the mixture of principal and other products actually produced by the sector. In this study purification of the paper industries is important because pollution is generated primarily in pulp mills (24.01). The inclusion of pulping processes in other paper industries (24.02–24.07) is more common than the inclusion of other paper processes within pulp mills. As a result, emissions are likely to be underestimated unless we work with clearly defined processes.

Input coefficients in an input-output matrix can be purified as follows. Let

A = $[a_{ij}]$ be a matrix of pure product coefficients;
P = $[p_{ij}]$ be the product mix matrix. It shows the value of product i produced by industry j in a given year; and
X = $[x_{ij}]$ be the (observed) input-output flow matrix. It shows the amounts of inputs i required to produce the actual product mix of sector j for each j.

Algebraically

$x_{ij} = a_{i1}p_{ij} + a_{i2}p_{2j} + ... + a_{in}p_{nj}$ in (i,j = i, ... n)

or $X = AP$

therefore $A = XP^{-1}$ (1)

Equation (1) gives a solution for the pure product coefficients A, given the observed flows X and the production matrix P.

The product-mix matrix [P] is a product-by-industry matrix. Each industry's column lists the quantity of each product that the industry produced. This matrix is obtained by transposing the matrix of secondary production and introducing the principal product output of each industry as the diagonal entry. Because industries are defined according to the principal-product rule, the diagonal entry will always be the largest entry in any column.

Equation 1 has been applied to economy-wide input-output models by Richard Stone [58] and others [3, 42]. The major problem associated with the implementation is the appearance of coefficients in the A matrix which defy common sense. Especially bothersome are the resulting negative coefficients. The method outlined in equation 1 can also be applied to a part of an input-output model. Since P has to be inverted, it must be a square matrix. X and A do not need to assume any particular shape, but the number of columns or industries to be purified has to equal the number of rows in P. To be meaningful, the product mix matrix P should contain all the products that any one of the industries listed in P produces. Since P must also be square, it is sometimes difficult to restrict its size.

To illustrate the point, let us assume that two industry input vectors need to be purified. Over 50 percent of the output of industry 1 is in product 1; over 50 percent of the output of industry 2 is in product 2. The only other product that industry 1 produces is product 2; the only other product that industry 2 produces is product 1. The P matrix can now be restricted to a two-by-two matrix that lists the quantities of each product that each industry produces. The outputs of the two industries consist entirely of the two products. If industry 2 produces a significant quantity of a third product, however, this third industry must be included in the P matrix. If this third industry produces quantities of products 4, 5, and 6 in addition to products 1, 2, and 3, then the number of sectors to be included in the P matrix gets much larger.

When only a limited number of related industries are purified, it is relatively simple to spot and eliminate irregularities in the results. Special knowledge of the industry can be brought to bear on the problems, and the methods of data collection for the industries can be scrutinized. The successful application of equation 1 on a limited scale hinges upon the existence of industry groupings that are self-contained with regard to products. Thus it is possible to purify input structures of groups of industries that produce each other's products and virtually no others. This proved to be the case for the paper industries analyzed in this report. Examination of product detail for the seven pulp, paper, and paper-converting industries revealed that, with one exception, these industries produce each other's products almost exclusively. Table 1 summarizes the production of these industries within and outside the paper groups. With the exception of industry 24.07 (converted paper products n.e.c., except containers and boxes), the outputs of more than 97.7 percent of each of the individual industries fall within the group. It is a

Table 1. *Production of Secondary Products by Various Industries, 1963 (thousands of dollars)*

	Pulp 24.01	Paper 24.02	Paper-board 24.03
Gross domestic output	1,157,100	5,054,400	2,557,400
Secondary production (transfers out)	71,446	390,041	349,512
To other sectors in the pulp, paper and converted paper group	64,018	346,280	309,669
To sectors outside the pulp, paper and converted paper group	7,428	43,761	39,843
Percent secondary production of gross domestic output	.64	.87	1.56

Source: 1963 Office of Business Economics input-output tables.

matter of judgment whether 2.24 percent of "other" output produced by industry 24.04 (envelopes) is significant enough to expand the industry group. For the purpose of this report it is assumed that the residual of 2.24 percent is not significant.

Only slightly more than 80 percent of the output of industry 24.07 is classified as paper and allied products—that is to say, nearly 20 percent of its output falls outside the industry group. Since industry 24.07 is a catch-all for converted paper products not elsewhere classified, we cannot identify a single technology for it. Table 2 demonstrates that industries 24.01–24.06 are involved to a limited extent in the production of products of in-

Envelopes 24.04	Sanitary Paper 24.05	Wall and Building Paper 24.06	Converted Paper n.e.c. 24.07
362.900	1,001,800	353,900	3,615,400
27,467	67,267	12,267	905,851
19,365	51,617	6,579	185,274
8,111	15,650	6,167	720,577
2.24	1.56	1.74	19.93

dustry 24.07. It seems reasonable, therefore, to exclude industry 24.07 from the computations.

The remaining six paper and allied products industries can be grouped in sequence by production stages. The first step in the production of paper is pulp-making, i.e. industry 24.01. The finished pulp next enters the paper mills, paperboard mills, or wallpaper and building paper mills (24.02, 24.03, 24.06). Finally, a portion of the output of these industries is converted into envelopes (24.04) and sanitary paper products (24.05).

A majority of the plants in industries 24.02 and 24.03 and a good many of the plants in industry 24.06 are integrated back-

Table 2. *Secondary Production by Other Paper and Allied Products Industries of Products Primary to Converted Paper Products n.e.c. (24.07) (thousands of dollars)*

Producing Sector	Pulp 24.01	Paper 24.02	Paper-board 24.03
Gross domestic output	1,157,100	5,054,400	2,557,400
Production of products primary to 24.07			
Value	—	28,386	17,745
Percent of gross domestic output	—	.56	.69

Source: 1963 Office of Business Economics input-output tables.

ward into pulp production. The pulp production of integrated facilities is not treated as a pulp input into the industry under which the facility is classified in the published tables. This becomes evident upon the examination of available technological data. In the case of paper and paperboard production, the pulp input amounts to nearly 50 percent of the value of the finished product. For construction board the coefficient is smaller but still more than 25 percent of the value of output. In contrast, the published pulp input coefficients for industries 24.02, 24.03, and 24.06 are .1189, .0569, and .0378 respectively [62]. A major portion of the pulp production in the economy is never recorded in the input-output tables, because it is produced in integrated paper facilities. The inputs necessary to produce this pulp, however, are recorded in the input vector of the industry under which the integrated facility is classified. The in-house production of pulp by the six paper and allied products industries has to be estimated if the purification computations are to be meaningful.

Envelopes 24.04	Sanitary Paper 24.05	Wall and Building Paper 24.06
362,900	1,001,800	3,539,000
19,012	51,526	537
5.24	5.14	.02

The reverse problem exists in the forward-integrated pulp mills. Some pulp facilities manufacture a limited amount of paper and paperboard. The pulp required to manufacture this paper and paperboard has not been entered in the input-output table, but for the computations to be meaningful it is necessary to estimate the amount of pulp. From the 1963 Census of Manufactures woodpulp purchases in tons of wood pulp and in dollar value can be obtained for industries 24.02, 24.03, and 24.06. An average price per ton of wood pulp can then be arrived at through division. The Census also records the quantity of in-house wood pulp produced and used by these industries. When the average price per ton of wood pulp is applied to these quantities, an estimate of the dollar amount of in-house wood pulp produced and used by industries 24.02, 24.03, and 24.06 is obtained. Table 3 demonstrates the computation.

A different approach must be used to compute the amount of pulp used in the production of paper and paperboard by forward-integrated pulp mills. The Census of Manufactures does

Table 3. *Wood Pulp Production by Backward Integrated Mills for In-House Use*

	Paper 24.02	Paper-board 24.03	Wall and Building Paper 24.06
Purchased wood pulp			
(1) tons	4,602,967	720,632	46,476
(2) thousands of dollars	$556,588	$67,850	$3,641
(3) Price per ton (2) ÷ (1)	$120.92	$94.15	$78.34
Own wood pulp produced			
(4) tons	10,323,776	11,307,570	1,130,453
(5) thousands of dollars (4) × (3)	$1,248,345	$1,069,647	$88,561

Source: [64].

not record in-house use of pulp by pulp mills. Since the amount of paper and paperboard produced by pulp mills is minor, it is possible to execute the pure product computations on a first round while including the figures of line (5) in Table 3. The resulting estimates of pulp input coefficients are then applied to the amounts of paper and paperboard produced to compute the pulp requirements of this paper and paperboard. Table 4 summarizes the computation. The computations are then repeated, this time including adjustments for pulp requirements of paper and paperboard produced by pulp mills.

The pure product vectors are given in Table 5. Where possible, the entries were verified against other industry and technical data. The estimates conform very well to the real world. Although

Table 4. *Pulp Requirements of Paper (24.02) and Paperboard (24.03) Produced by Pulp Mills (thousands of dollars)*

	Paper 24.02	Paperboard 24.03
(1) Production of pulp mills	$23,081	$41,937
(2) "Interim" pure product pulp coefficient	.4937	.5337
Pulp required (1) × (2)	$22,382	$11,395

a few small negative values appeared in the first round of pure product coefficient estimates, they were negligible. In the vectors in Table 5 negative coefficients were eliminated and positive coefficients reduced proportionately by their sum.

3.2 Water Pollution Coefficients

Of the four most widely-used pulping processes—sulfite, sulfate, NSSC, and groundwood pulping—the sulfite process poses the biggest water pollution problem. The principal pollutants are dissolved substances like carbohydrates, salts, and soaps. The sulfate process is considerably cleaner for steams because most of the soluble organic material in the waste liquor is burned in the alkali recovery process. But, as noted earlier, the burning of this spent liquor causes considerable air pollution.

The mechanical process for separating the cellulose fibers, i.e. groundwood pulping, is the least polluting of the pulping processes. From it a semichemical process known as NSSC was developed in the early 1950's. Like sulfate and sulfite pulping, it causes extensive water pollution. The spent liquor of the NSSC process may be concentrated and burned to recover sodium sulfate, which is an input in the sulfate pulping process. Some

Table 5. *Pure Product Coefficients for the Paper and Allied Products Industries (dollars per ten thousand dollars of output)*

No. Supplying Sector	Pulp 24.01	Paper 24.02	Paper-board 24.03	Envelopes 24.04	Sanitary Paper 24.05	Wall and Building Paper 24.06	Converted Paper n.e.c. 24.07
7 Coal mining	37	87	111	–	1	64	6
9 Stone & clay min. & quar.	7	97	3	–	–	329	11
10 Chem. & fert. mineral min.	63	–	–	–	–	–	–
12 Maint. & rep. constr.	255	–	–	16	19	–	18
14 Food & kindred prod.	98	124	36	286	50	118	52
16 Br. & nar. fabrics, yarn, and thread mills	–	152	–	–	–	2	47
17 Misc. tex. goods & floor cov.	–	–	–	–	179	15	–
18 Apparel	6	6	4	14	5	10	9
20 Lumber & wood, exc. cont.	2175	170	238	4	5	207	8
24.01 Pulp	86	4694	4838	–	4231	3150	31
24.02 Paper	9	5	6	2946	38	236	2663
24.03 Paperboard	9	–	–	–	38	–	460
24.04 Envelopes	1	1	1	5	1	1	1
24.05 Sanitary paper	1	1	1	3	6	2	2
24.06 Wall & bldg. paper	–	–	–	–	–	10	–
24.07 Converted paper n.e.c.	–	–	–	–	–	–	22
25 Paperboard cont. & boxes	–	123	–	225	976	39	595
26 Printing and publishing	3	–	–	72	2	–	83
27 Chemicals & sel. chem. prod.	813	211	116	239	–	65	242
28 Plastics & synthetic mat.	35	77	65	–	–	44	348
29 Drugs, clean. & toilet prep.	18	13	8	1	–	8	1
31 Petro. ref. & rel. indust.	131	33	77	12	20	135	84
32 Rubber and misc. plastic	–	19	19	255	32	20	380

33 Leather tan. & ind'l. prod.	4	2	1	—	—	4	1
36 Stone & clay products	33	12	—	—	—	—	—
37 Prim. iron & steel manuf.	—	2	—	—	—	—	13
42 Other fabric. metal prod.	432	99	96	13	14	53	2
46 Mat. handl. mach. & equip.	8	4	1	2	—	8	6
48 Spec. indust. mach. & equip.	162	—	—	—	—	—	2
49 Gen. indust. mach. & equip.	74	14	—	—	—	6	1
59 Motor veh. & equip.	—	—	1	1	1	1	2
62 Scient. & contr. instr.	1	1	1	3	1	2	1
63 Optical, opthalmic and photographic equipment	—	—	—	1	—	1	1
64 Misc. manufacturing	1	1	—	2	1	1	1
65 Transport. & warehousing	501	312	247	216	271	175	249
66 Commun. exc. radio & TV	22	24	14	60	19	29	43
68 Elec., gas, water, sanit. serv.	504	177	42	70	30	382	79
69 Wholesale & retail trade	307	185	348	284	369	274	334
70 Finance & insurance	51	56	36	51	16	72	33
71 Real estate & rental	59	12	1	150	76	28	134
72 Hotels, personal & repair services, exc. auto	4	3	2	36	36	6	29
73 Business services	118	156	104	169	135	210	171
75 Auto repair & services	5	5	10	10	4	12	8
77 Medic., educ., & non-prof. org.	3	3	2	8	2	4	6
78 Fed. gov't. enterprises	9	7	8	13	5	8	10
79 State & local gov't. ent.	7	5	4	1	—	3	1
81 Bus. travel, entert., gifts	15	11	8	126	126	24	108
82 Office supplies	6	6	3	11	4	7	8
83 Scrap, used, 2nd-hand goods	—	149	719	—	—	429	14
Value added	3928	2938	2832	4694	3325	3808	3681
Total	10000	10000	10000	10000	10000	10000	10000

Table 6. *Water Pollution by the Four Most Widely Used Pulping Processes (pounds per ton of output)*

	Emissions		
Process	Pounds BOD$_5$ Per ton of Output	Suspended Solids	Weights Based on 1961 Production
Groundwood			.1361
old	107	84	.0327
present	36	20	.0925
new	20	24	.0109
Sulfate			.6547
old	156	94	.1571
present	95	105	.4452
new	37	22	.0524
Sulfite			.1091
old	489	151	.0262
present	289	69	.0742
new	88	22	.0087
NSSC			.1001
old	333	210	.0261
present	69	40	.0740
Totals			1.0000
			1.0000

Average emissions: 125 pounds of BOD$_5$ per ton of output.
                     85 pounds of suspended solids per ton of output.

Source: [65].

companies have established sulfate mills and NSSC mills on the same site to take advantage of this cross-cycle possibility.

Discharges of BOD and suspended solids per ton of pulp

Table 7. *Water Pollution from Paper and Paperboard*
Mills and from De-Inking
(pounds per ton)

	BOD_5		Suspended Solids	
	Range	Estimate Point	Range	Estimate Point
Paper mills	10–20	16	30–60	48
Paperboard Mills	35	35	40	40
Newsprint de-inking	20–60	40	25–220	123
High grade de-inking	33–134	84	600–1100	850

Sources: [2, 13, 14].

produced are given in Table 6. On the average, 125 pounds of BOD_5 are produced per ton of pulp in the United States. This average is the sum of point estimates of BOD for the four pulping processes and two or three other technologies, weighted by the 1961 output distribution (column 3, Table 6) and adjusted for output not accounted for by the four processes. The differences in BOD_5 emission per ton of output among the pulping processes are sizable. The average rate of emission is, therefore, very sensitive to the pulping mix. A one percent shift of production from sulfate pulping to sulfite pulping, for example, would increase the average pounds of BOD_5 per ton by 2.15 pounds BOD_5, or 1.72 percent.

The extent of water pollution in terms of suspended solids for each of the four pulping processes is also shown in Table 6. The suspended solids pollution varies less among the different processes. As a result, the average amount of suspended solids per ton of pulp produced in the United States is less sensitive than BOD_5 to the pulp mix. Similar water pollution statistics for paper mills, the paperboard mills (including construction board), news-

Table 8. *Water Pollution Coefficients*
(Physical and 1963 Dollar Units)

	Pulp 24.01	Paper 24.02	Paper-board 24.03	Wall and Build-ing Board 24.06	News-print de-inking	High Grade de-inking
1. pounds BOD$_5$ per ton of output	125	16	35	35	40	84
2. pounds suspended solids per ton of output	85	48	40	40	123	850
3. price per ton of output	118.-	225.-	128.-	94.-	53.-	114.-
4. pounds BOD$_5$ per dollar of output (1) ÷ (3)	1.060	.073	.273	.372	.755	.737
5. pounds suspended solids per dollar of output (2) ÷ (3)	.360	.213	.313	.426	2.321	7.456

Sources: [2, 13, 14, 64].

print de-inking facilities, and high grade paper de-inking plants are presented in Table 7. The information on water pollution per ton of output by the various pulp and paper facilities can be combined with information on the price per ton of output from the Census of Manufactures [64] to give an estimate of pollution per dollar of output (numbers 4 and 5, Table 8). These are the water pollution coefficients incorporated in the input-output analysis.

3.3 Air Pollution

The sulfate (Kraft or Alkaline) process is the only one of the four major pulping methods that releases substantial quantities

Table 9. *Pounds of Air Pollution per Ton of Output in Sulfate Pulping*

	Emissions					
Pulping Stage	Partic- ulates	Hydro- gen Sulfide	Methyl Mercaptan	Dy- methyl Sulphide	Sulphur Dioxide	Carbon Monoxide
Digester blow system		.45	2.5	1.37		
Smelt tank	20 5 2	.03				
Lime kiln	18.7	1				
Recovery furnance	150 28 21	3.6 .7	5	3	5	60
Total						
Low	31.7	2.18	7.5	4.37	5	60
High	188.7	5.08	7.5	4.37	5	60

Source: [20].

of air pollutants. Unfortunately, nearly 60 percent of the pulp in the United States is produced by this method. Table 9 shows the major sources of pollutants in the sulfate process. Because it appears from the literature that most mills have the equipment required to operate with the low-level pollution coefficients, the low estimates are used in this study. The coefficients in Table 9 are converted from pounds-per-ton of sulfate pulp to pounds-per-dollar by division by the average sulfate pulp price. The final coefficient is weighted by .582 to account for the fact that only 58.2 percent of the pulp output in the United States is sulfate pulp. Table 10 gives the coefficients used in the input-output analysis. The pollution emitted per ton of pulp produced depends on the designed capacity of a pulp mill and the level of operation. The coefficients are relatively stable up to design capacity. Beyond design capacity, they increase rapidly. Total mill pol-

Table 10. *Air Pollution Coefficients for the Pulp Sector*

		Partic- ulates	Hydro- gen Sulfide H_2S	Methyl Mercap- tan RSH	Dy- methyl Sulphide RSR	Sulphur Dioxide SO_2	Carbon Monoxide CO
Pounds per ton of sulfate pulp	(1)	31.7	2.18	7.5	4.37	5	6.
Pounds per dollar of sulfate pulp (1) ÷ 150	(2)	.2113	.0145	.0500	.0291	.0333	.4000
Pounds per dollar of total pulp production (2) × .582		.1230	.0084	.0291	.0169	.0194	.2328

lutants emitted will therefore depend on mill output. The exact effect of mill operation beyond design capacity on the functioning of abatement equipment must be explored more thoroughly, since it appears from the literature that this is a common phenomenon in the industry. Apparently, mill management's performance is measured by how far it can "push" the mill. The problem is especially important because the composition of pollutants as well as their quantity may increase disproportionately with the quantity of pulp produced beyond design capacity.

3.4 Input Coefficients for the Paper Recycling Industries

There are five distinct paper recycling activities: collection, sorting, cleaning, de-inking, and repulping. Collection and sorting are very labor-intensive activities. The data examined for these industries indicate that more than 85 percent of collection costs are attributable to wages and salaries. The figures for sorting are similar. Since nonlabor costs of collection and sorting are

Table 11. *Input Coefficients for De-Inking and Repulping of Wastepaper*

No. Input	Newsprint De-inking Dollars per ton	Newsprint De-inking Dollars per Dollar	High Grade Groundwood Free Pulp Dollars per ton	High Grade Groundwood Free Pulp Dollars per Dollar
Collection	$33.30	.6283	$79.30	.6956
Wastepaper	30.00		62.60	
Processing Loss	3.30		15.70	
27 Chemicals	3.50	.0660	4.50	.0395
68 Utilities	3.50	.0660	6.30	.0553
Steam and Power	3.50		3.60	
Water	—		2.70	
Value added	12.70	.2396	23.90	.2096
Labor	3.80		3.50	
Depreciation	3.00		7.00	
Maintenance	—		4.00	
Supervision	—		1.35	
Miscellaneous	5.90		9.05	
Total	$53.00	1.0000	$114.00	1.0000

Source: Mr. Frank Hamilton at Charles Maine Inc., Boston, Mass.

primarily capital costs, all costs for these activities are value added. Operational data on collecting and sorting were obtained from the town of Hempstead, New York [50]. The town collects ten tons a day with a compactor truck operated four days a week by two collectors and a driver. From these figures we calculated that it takes 150 man years to collect 100,000 tons of waste paper. To account for sorting, baling, and supervision for newspaper and paper recycling, the figure was doubled. Recycling for paperboard and construction board manufacturing is less labor intensive. Virtually no sorting takes place, and collection is from office buildings and stores rather than residences. The employment coefficient was estimated as 150 man years for 100,000 tons. Technological coefficients for the cleaning, de-inking, and repulping activities were obtained from engineering sources. The coefficients computed in Table 11 are those used in the input-output model.

3.5 Abatement Coefficients

Operating and capital costs for treatment of pulp and paper mill effluent were obtained from *The Cost of Clean Water* [65]. These costs are presented for the effluents of two different processes (sulfite and sulfate pulping), and three different technologies (old, present, and new). The resulting six operating and capital cost vectors are weighted by the relative contributions of the corresponding processes and technologies to the combined output. Subsequently they are aggregated into an average operating and an average capital-cost vector for water pollution abatement per ton of air dry pulp.

The input-output computations require abatement vectors per unit of pollutant abated. To establish the quantity of BOD_5 abated for each ton of pulp output, the effectiveness of BOD_5 abatement has to be estimated. The abatement vectors per ton of pulp are assumed to remove 85 percent of the BOD_5 emitted by an average ton of pulp. The resulting pollution abatement vectors are presented in Table 12. Certain important assumptions are made about the vectors in this table. Although they are computed from cost estimates for water pollution abatement in only two pulping processes, it is assumed that input requirements per pound of BOD_5 abated are the same across the entire spectrum of pulp and paper industries. This assumption is not unreasonable. The technology for BOD_5 abatement is fairly uniform; variations are based on plant location and space requirements of the different abatement techniques rather than on specific product. Input-output analysis generally assumes that input requirements per unit of output—in this case per unit of BOD_5 abatement—are constant. From the literature on pollution abatement it is evident that abatement costs rise more than proportionally to the percentage of pollution abated. One way

Table 12. *Abatement Costs per Ton BOD$_5$ Eliminated*

Supplying Sectors		Dollars per Ton BOD
No.	*Current Account*	
12	Maintenance and repair construction	$ .77
27	Chemicals and selected chemical products	19.06
42	Other fabricated metal products	.44
49	General industrial mach. and equipment	2.98
68	Electric, gas, water and sanitary services	6.50
VA	Value added	18.70
	Capital Account	
11	New construction	83.41
42	Other fabricated metal products	30.38
49	General industrial machinery and equipment	68.53
53	Electric industrial equipment and apparatus	11.38

to make the model more realistic is to treat different portions of emissions of a given pollutant by a single industry as separate pollutants. If the last 5 percent of pollution emitted by a certain industrial process can be eliminated only by an abatement process that is technologically different from the abatement process used for the first 95 percent of the pollutant (i.e. requires more of other inputs per unit of pollutant abated), it can be routed through a different abatement process and treated on the model as a different pollutant. Lack of sufficient data precludes such an approach at present. As more data become available, it may be possible to refine the computations in this way.

The air pollution abatement vectors were obtained from research carried on by Terry Jenkins at the Harvard Economic Research Project. These vectors only apply to particulate emissions. As indicated earlier, the abatement of gaseous emissions

Table 13. *Abatement Costs per Ton of Particulates Eliminated*

Supplying Sectors		Dollars per Ton
No.	*Current Account*	
12	Maintenance and repair construction	.376
68	Electric, gas, water and sanitary services	.685
VA	Labor, capital — 10 year life	10.375
	Capital Account	
11	New construction	1.638
40	Heating, plumbing and structural metal products	23.433
49	General industrial machinery and equipment	19.875
55	Electric lighting and wiring equipment	4.082
62	Scientific and controlling instruments	2.028
65	Transportation and warehousing	.671
73	Business services	20.385
VA	In-house engineering and labor	26.267

is still very much in the experimental stage and no standard technology has yet been developed. Gaseous emissions are not abated in the model. The operating and capital cost coefficients are presented in Table 13. Similar reservations apply to the use of these vectors as discussed for water pollution abatement.

3.6 Employment Coefficients

Employment coefficients for the 83-industry classification used by the Office of Business Economics were assembled at the Harvard Economic Research Project from data in the Census of Manufactures. The employment coefficients for the seven detailed paper and allied products industries were computed from the 1963 *Census of Manufactures*. Because six of the detailed paper and allied products industries were computed on a pure product basis, it was necessary to purify the employment coefficients in a similar way.

D_j is the observed employment in industry j. E_i is the employment per unit of product i produced. P_{ij} is the quantity of product i produced by industry j. For each of the products produced by industry i, the number of workers employed by the industry is proportional to the amount of that product produced. Hence

$$D_j = E_1 p_{1j}, + E_2 p_{2j}, + \ldots \ldots + E_n p_{n,j} \ (i = 1, 2, \ldots \ldots, n)$$
or $D' = EP$ in which $P = [p_{ij}]$ $\qquad\qquad$ (2)
$\qquad E' = DP^{-1}$

where D' is a (row) vector of industry employment coefficients and E' a (row) vector of industry product coefficients. The matrix P^{-1} has already been computed. When it is applied to the observed employment coefficients D', the result is a vector of employment E' by pure product industries.

Employment coefficients for the recycling industries were obtained from various sources. For the two de-inking industries employment coefficients were provided by Frank Hamilton of Charles Main, Inc., Boston, Massachusetts.

4.0 THE IMPACT OF RECYCLING

4.1 Limits to Recycling

The data in the previous pages make it possible to estimate the economic and environmental impacts of an increase in the recycling of wastepaper. Given that a certain quantity of paper and paperboard will be produced by the economy in a particular year, it will be useful to know how the economy and the environment are affected when a portion of the pulp requirement is satisfied by secondary fiber pulp rather than virgin fiber pulp.

An understanding of the effect of such a shift on net emissions, outputs, and employment will give some systematic basis for judging the value of recycling wastepaper.

It has been widely accepted that an increase in the use of wastepaper is feasible in the United States. Engineering sources confirm that it is technically possible for large amounts of waste-paper to be recycled without impairing paper quality and with-out increasing the price of the end product [45]. In 1947 waste-paper constituted 35.2 percent of all fibrous inputs into paperboard pulp manufacturing. By 1963, however, the per-centage had dwindled to 23.4 percent; by 1970, to 22.0 percent [8]. For the purpose of this study the marginal quantity by which the amount of recycled wastepaper is to be increased was established at 100,000 tons. Alternatively, it would have been possible to compute the net economic and environmental impact of a one percent increase in the rate of wastepaper re-cycling. Since the output of paper and paperboard increases steadily, the denominator of a percentage figure would change over time. Thus the estimated impact would be valid only for a base year. The advantage of using a number of tons of recycled wastepaper (replacing a corresponding amount of virgin fiber pulp) is that the net result can be extended to any amount of pulp wastepaper recycled, by simple scalar multiplication.

The extension of the results to very large amounts of recycled waste paper is limited, however, by known nonlinearities, and by such other constraints in the recycling industries as the quantity of recycled paper increases. For example, collection may become disproportionately more expensive as increasing amounts of wastepaper are recycled. (The experience of the Garden State Paper Company has shown, however, that with proper organiza-tion the amount of wastepaper collected from households can be vastly increased without an increase in the cost of collection. Greater efficiency may well come with more stable markets for wastepaper.) There is, of course, a limit to the amount of waste-paper that can be recycled. From the experience of other indus-

trial nations, such as Japan and West Germany, it would seem
that a recycling rate of 35–40 percent is possible. By computing
1947 United States rates of recycling wastepaper into various
grades of paper, taking into account additional technological
information on the improved feasibility of newspaper recycling,
we estimated that a 40.9 percent rate of recycling would have
been feasible in 1963. The total amount of wastepaper recycled
would then have been 16.8 million tons. In fact, however, 9.6
million tons were recycled. In other words, an additional 7.2
million tons of wastepaper might have been recycled without
impairing paper quality or increasing prices. On the same basis,
the 12 million tons of wastepaper recycled in 1970 could have
been increased by 9.8 million tons.

The above estimates establish some limits within which the
results of our impact study may be valid. We now consider the
impact of a shift from the use of virgin fiber pulp a to secondary
fiber pulp on the economy and the environment.

4.2 Pollution Trade-offs in Recycling

Table 14 shows the computed impact on the environment of
recycling 100,000 tons of wastepaper into four different grades
of paper and paperboard. The negative numbers indicate a net
reduction in the quantity of the specific pollutants emitted; the
positive numbers indicate a net increase. According to the table,
there is a decrease in emissions of most pollutants where there is
an increase in the recycling of paper. Only two positive estimates
appear in the exhibit. When newsprint is manufactured from
wastepaper rather than from virgin fiber woodpulp, emissions
of suspended solids increase substantially. Similarly, in paper
manufacturing the use of recycled wastepaper produces more
suspended solids than does virgin fiber pulp. For all other pollutants, however, a net decrease occurs when paper is manufactured from recycled stock rather than virgin fiber pulp.

Table 14. *Environmental Impact of Substituting 100,000 Tons of Wastepaper for Equivalent Pulp in the Manufacture of Paper and Paperboard*

	Total Emissions for Paper and Allied Products with 1963 Output	Change in Effluent Discharge When the Substitution takes Place in			
		Newsprint	Paper	Paperboard	Constr. Board
Water Pollution					
tons BOD_5	2,464,397	-1,109	-2,611	-3,169	-1,325
tons suspended particles	2,168,090	+3,197	+26,250	-2,154	-901
Air Pollution					
tons ambient particles	226,357	-320	-647	-367	-154
Tons H_2S	15,459	-22	-44	25	-10
Tons RSH (methyl mercaptan)	53,738	-76	-159	-87	-36
Tons RSR (dimethyl sulphide)	31,286	-44	-89	-51	-21
Tons SO_2	35,702	-50	-102	-58	-24
Tons CO	428,427	-606	-1,225	-695	-291
Solid Waste (tons)	—	-100,000	-100,000	-100,000	-100,000

4.3 Abatement and Disposal Costs Associated with Recycling

The costs of total pollution abatement and solid waste disposal can be used to estimate the first-round effect of an increase in recycling of wastepaper. By weighting Table 14 estimates by the cost of eliminating pollutants, we can gain some insight into the relative desirability of recycling paper. BOD_5 and suspended solids pollution are abated by the same water treatment. An increase in either, therefore, means an increase in the cost of water pollution abatement. The cost of abatement of suspended solids is estimated at $71.26 per ton. The cost of abatement of ambient particles is estimated at $11.44 per ton. The cost of solid waste collection and disposal varies considerably from one region to the other. In the city of New York the average cost per ton of refuse collected is $27.28; disposal by landfill or incineration is $3.78 per ton [23]. In the city of Madison, Wisconsin, costs range from $4.25 to $38.17 per ton collected. In Cambridge, Massachusetts, the per ton cost for solid waste management is estimated at $30.00 [11]. For this study an average cost of $20.00 per ton for collection and disposal seems reasonable. No abatement costs for the gaseous emissions are available. Although Table 14 shows that increased recycling decreases the quantity of gaseous emissions, this is not taken into account in the computation of the net effect of recycling.

Table 15 demonstrates that when the quantity of recycled paper is increased by 100,000 tons per year, air pollution abatement and solid waste disposal are greater than the increased costs for water pollution abatement. In terms of environmental protection, then, recycling wastepaper is more economical than producing paper from virgin fiber pulp. Table 15 does not attempt to quantify the harmful effects of different pollutants upon the environment. Instead, it shows that if full abatement and disposal policies were implemented for the three classes of pollutants listed, wastepaper recycling would make possible a net reduction of abatement expenditures. It is difficult to say

Table 15. *Net Costs of Increasing Wastepaper*
Recycling by 100,000 Tons
(dollars)

	Substitution of Recycled for Virgin Pulp in			
	News-print	Paper	Paper-board	Constr. Board
Water pollution abatement cost	+279,125	1,870,575	−153,494	−64,205
Air pollution abatement cost	−3,661	−7,402	−4,198	−1,762
Solid waste disposal cost	−2,000,000	−2,000,000	−2,000,000	−2,000,000
Net effect − net saving + net cost	−1,729,536	−136,827	−2,157,692	−2,065,967

who will benefit from this net saving, however. Since water pollution and air pollution abatement will probably be the responsibility of the paper industry, a reduction in these costs will benefit the industry. A reduction in solid waste disposal expenditures will probably benefit individual communities since they ordinarily take charge of the collection and disposal of solid waste. This will be especially true when residential wastepaper is recycled.

Table 16 lists the industries which would be most seriously affected by an increase in paper recycling. The negative numbers indicate a reduction in gross output; the positive numbers indicate an increase in the industry's gross output. A shift from virgin fiber pulp to recycled paper would mean less demand on most of the industries, except those directly involved in recycling. In other words, recycling paper means that less of the economy's industrial resources is needed. The negative entries in Table 16 are exaggerated. The input structure of the de-inking industries is obtained from engineering sources and does not include inputs that only become apparent when a great number of operating data from various sources are combined. The relatively sparse

Table 16. *Changes in Sectoral Output Levels with Substitution of 100,000 Tons of Wastepaper for Equivalent Pulp in the Manufacture of Paper and Paperboard (thousands of dollars)*

			Change in Output When the Substitution Takes Place in			
No.	Industry	1963 GDO	Newsprint	Paper	Paperboard	Constr. Board
3	Forestry and fishery products	1,287,000	-152	-309	-176	-73
12	Maintenance and repair construction	19,861,000	-176	-368	-240	-96
20	Lumber and wood products, except containers	9,974,000	-1,582	-3,200	-1,818	-757
24.01	Pulp mills	3,681,000	-5,202	-10,525	-5,974	-2,497
	Newsprint de-inking	n.a.	4,370	-0-	-0-	-0-
	Paper de-inking	n.a.	-0-	8,060	-0-	-0-
	Collection and sorting	n.a.	+2,746	-5,607	2,275	1,378
27	Chemicals and selected chemical products	16,591,000	-214	-754	-656	-274
31	Petroleum refining and related industries	21,028,000	-128	-304	-192	-96
37	Primary iron and steel	23,869,000	-144	-272	-176	-80
42	Other fabricated metal	8,830,000	-271	-548	-315	-130
68	Electric, gas, water, and sanitary services	29,649,000	-332	-224	-464	-192

vector of input coefficients from the de-inking industries replaces in part the relatively full vector of coefficients from the pulp industry. As a result, the predicted decline in the industrial demand for products is overstated.

A similar word of caution is in order for the net effect on employment in Table 17. Because input coefficient vectors for de-inking and collection may be incomplete, employment requirements for increased recycling may be underestimated. As expected, the adverse economic impact would fall primarily on the pulp mills and the lumber and wood products industries. A more subtle shift, not directly apparent from the figures, would also take place. Most pulp mills are part of integrated pulp and paper facilities. These integrated plants are located predominantly in the rural areas of the Northeast, the South, and the Northwest. Collection, de-inking, and recycling are urban activities. Most de-inking plants are found near the large population centers of Boston–New York–Washington, Chicago–Pittsburgh, and San Diego–Los Angeles. It is unlikely that secondary fiber pulp manufactured in the urban centers will be dried and shipped to rural paper plants, there to be repulped for processing and returned to the major paper markets in the urban centers. Integrated de-inking and papermaking facilities are more likely to be located in the urban centers, nearer to materials and markets. Only one company in the United States (Kimberly Clark) actually sells pulp manufactured from recycled paper. All of its customers use the pulp at facilities that are no more than ten miles away from the manufacturing site. This suggests that the large negative estimates shown in Tables 16 and 17 for output and employment in the pulp, lumber, and wood industries apply to rural areas, whereas the positive estimates apply to the urban areas.

By and large, reductions in output will take place in industries characterized by a sophisticated technologies. Pulp mills require high engineering skill input. So do the chemical and petroleum refining industries. Collection and sorting, on the other hand, are labor-intensive industries with relatively simple skill requirements. Therefore, an increase in recycling would not simply

Table 17. *Major Changes in Employment with Substitution of 100,000 Tons of Wastepaper for Equivalent Pulp in the Manufacture of Paper and Paperboard*

No.	Industry	1963 Employment	Change in Employment When the Substitution Takes Place in			
			Newsprint	Paper	Paperboard	Constr. Board
3	Forestry and fishery products	168,904	-20	-41	-23	-10
12	Maintenance and repair construction	1,198,522	-11	-22	-14	-6
20	Lumber and wood products, except containers	691,064	-110	-222	-126	-52
24.01	Pulp mills	90,912	-128	-260	-148	-62
	Newsprint de-inking	—	37	—	—	—
	Paper de-inking	—	—	82	—	—
	Collection and sorting	—	300	300	150	150
27	Chemicals and selected chemical products	464,301	-6	-21	-18	-8
31	Petroleum refining and related industries	196,964	-1	-3	-2	-1
37	Primary iron and steel manufacturing	885,723	-5	-10	-7	-3
42	Other fabricated metal prod.	485,241	-15	-30	-17	-7
68	Electric, gas, water, and sanitary services	685,827	-1	-5	-11	-4
	Total (does not equal column sums because of omissions)		-145	-597	-447	-87

cause a net reduction in the number of jobs. It would also mean an increase in urban areas of jobs with simple skill requirements and a loss in rural areas of jobs with higher skill requirements.

5.0 THE IMPACT OF POLLUTION ABATEMENT

5.1 Economic Impact

The data described in Section 3 permit us to estimate the impact of increased pollution abatement in the paper and pulp industries. The estimates refer to BOD_5 abatement and the removal of ambient particles. Abatement of suspended solids takes place concurrently with BOD_5 removal in the treatment of mill effluent. Thus, although only BOD_5 will be mentioned, the abatement of suspended particles is simultaneously accounted for. Since there are no accepted techniques for abating gaseous emissions, the analysis in this section does not include an estimate of the impact on the economy of abatement of gases.

In 1968 approximately 90 percent of particulate emissions were abated [68]. In 1963 approximately 60 percent of suspended solids were removed from bodies of water and 28 percent of the BOD_5 load of pulp and paper mill effluent was eliminated prior to discharge. On the basis of the pollution coefficients described in Section 3.4, the total BOD_5 load discharged in 1963 was estimated at 2154 thousand tons, not very different from the 2950 tons cited in *The Cost of Clean Water* [65].

The emissions of ambient particles by the pulp and paper industries in the same year are estimated at 238 tons. These figures were compared to those cited in other sources [67, 68] and were found to be of a similar order of magnitude. Vandegrift [68] arrives at a quantity of 373 tons of particulate emissions. This figure is necessarily overestimated because the quantity of Kraft pulp produced is overestimated by approximately 50 percent. Adjusting for this error would reduce Vandegrift's estimate to 249 tons of particulate emissions, very close to the estimate of 238 tons made in this study.

For this analysis, costs of abating BOD_5 and ambient particle pollutants are divided into four categories. The first category consists of the initial capital investment necessary to abate the pollutants that are still emitted with the 1963 levels of control. This initial capital investment is a one-time expenditure. It is difficult to predict its timing. Because of the annual growth of the paper industry, the proportionate size of this initial investment depends upon the year in which it takes place. Table 18 lists the lump sums of investment needed to bring pollution control in the pulp and paper industries up to specific control levels (85 percent for BOD_5 and 99.9 percent for ambient particles).

Capital expenditures in 1963 in the paper and allied products industry amounted to $732 million. They rose to $1489 million in 1967. The initial pollution abatement capital investments necessary are therefore substantial. If all clean-up had taken place in a single year (1963), capital expenditures would have had to exceed actual investment by 56.7 percent for the abatement of BOD_5 and by an additional 2.6 percent to control ambient particle pollution effectively. Presumably, the investment in abatement equipment would be spread over several years.

Once the initial investment has been made, the cost to keep the pollution abatement control levels at 85 percent for BOD_5 and 99.9 percent for ambient particles can be divided into three categories: (1) operating costs of the additional abatement equipment; (2) replacement investment to cover the attrition of the new equipment; and (3) additional investment for abatement because of expansion of the pulp and paper industry. These costs are all based on differences from 1963 industry practice. On the average, the industry abated approximately 90 percent of the particulate air emissions and 28 percent of the BOD_5 pollutant in 1963. It is technically possible to increase the abatement to 99.9 percent for ambient particles and to at

Table 18. *Initial Investment Necessary to Increase BOD₅ Abatement to 85 Percent and Particulates Abatement to 99.9 Percent* *(millions of dollars)*

No.	Supplying Sectors	Assuming output levels of		
		1963	1970	1980
	To abate BOD₅			
11	New construction	177	227	382
42	Other fabricated metal products	65	83	139
49	Gen'l. industrial mach. and equip.	148	186	314
53	Electric indust. equip. and apparatus	25	31	52
	Total initial investment to abate BOD₅	415	527	887
	To abate ambient particles			
11	New construction	—*	—	1
50	Machine shop products	4	6	9
59	Motor vehicles and equipment	4	5	8
65	Transportation and warehousing	1	1	2
72	Hotels; personal & repair exc. auto	—	—	1
75	Auto repair and services	—	—	—
83	Scrap, used, & secondhand goods	4	5	8
	Value added	5	6	11
	Total initial investment to abate ambient particles (does not equal column sums because of omissions)	19	23	40

*less than one million dollars.

least 85 percent for BOD₅. The operating costs, replacement investment, and additional investment for abatement in newly added capacity are all computed on the basis of the difference between average industry practice and this technically feasible industry practice. After the initial investment has been made,

the continuous effect of the maintenance of the new levels of
pollution abatement on the economy can be determined.

Table 19 lists the additional outputs needed from each supplying industry to sustain the new levels of pollution abatement attained in 1970, after the initial investment is made.
Only one industry—chemicals and selected chemical products
(27)—has an output that rises by more than $50 million. Gen-
eral industrial machinery and equipment (49) and electric, gas,
water, and sanitary services (68) fall in the $25–50 million
interval. These figures support the electric utilities' claim that
their industry will be a major force in cleaning up the environment. Construction (11), primary blast and steel manufacturing (37), and other fabricated metal products (42) fall in the
$10–25 million interval. The paper and allied products indus-
tries (24.01 to 24.07) will have to expand their outputs by
$3.4 million to abate their own pollution. Since the direct coef-
ficients for abatement do not include a coefficient for paper
and allied products, these abatement requirements must be
required indirectly. Column 1 (output devoted to operation)
shows that $2.3 million of the increase in the paper and allied
products industry output was indirectly required for operation
of the additional abatement equipment, rather than for the
replacement of this equipment or for expansion of the industry.

Table 20 is an identical table for 1980. In Table 21, the industries with an increase in output of more than $10 million in
Tables 18 and 19 are ranked in order of their relative increases.
Two observations are in order. (1) Between 1970 and 1980, the
number of industries whose output for additional pollution
abatement in pulp and paper passed the $10 million mark nearly
doubled. (2) Of the eleven industries listed for 1980, six are
direct suppliers to BOD_5 abatement or ambient particle control,
but the other five (31, 37, 38, 65, 69) are only indirectly affected
by the new pollution abatement practices. The value of a general
equilibrium model in analyzing the impact of pollution abatement is evident. These five industries would not have been identified in a partial analysis.

Table 19. *Changes in Output Levels to Maintain Improved Standards of Air and Water Pollution Abatement in the Paper Industries, 1970 (thousands of dollars)*

No. Industry	Opera-tion	Replace-ment	Expan-sion	Total
1 Livestock and livestock products	669	127	133	929
2 Other agricultural products	707	218	228	1153
3 Forestry and fishery products	210	114	120	444
4 Agric., forestry, fishery services	58	21	22	101
5 Iron & ferroalloy ores mining	574	249	259	1081
6 Nonferrous metal ores mining	715	249	259	1223
7 Coal mining	1209	171	177	1557
8 Crude petroleum and natural gas	4697	372	387	5457
9 Stone and clay mining & quarrying	321	241	254	817
10 Chemical & fertilizer mineral mining	2008	31	33	2072
11 New construction	—	11,378	12,040	23,418
12 Maintenance & repair construction	4843	345	355	5543
13 Ordinance and accessories	17	28	29	74
14 Food and kindred products	1923	262	273	2458
15 Tobacco manufactures	35	13	14	62
16 Br. & nar. fabrics, yarn, thread mills	207	130	132	469
17 Misc. textile goods and floor coverings	94	109	112	315
18 Apparel	99	57	59	216
19 Misc. fabricated textile products	91	29	26	146
20 Lumber & wood prod., exc. containers	877	1153	1216	3246
21 Wooden containers	38	19	20	77
22 Household furniture	12	78	82	172
23 Other furniture & fixtures	7	42	44	93
24.01 Pulp	530	121	127	778
24.02 Paper	498	126	131	754
24.03 Paperboard	441	94	98	632
24.04 Envelopes	39	12	12	63
24.05 Sanitary paper	22	11	11	44
24.06 Wall and building paper	43	29	31	103

No. Industry	Oper-ation	Replace-ment	Expan-sion	Total
24.07 Converted paper n.e.c.	737	125	131	994
25 Paperboard containers and boxes	678	197	205	1080
26 Printing and publishing	1100	413	430	1943
27 Chemicals and sel. chemical products	65,579	727	757	67,063
28 Plastics and synthetic materials	1724	221	229	2174
29 Drugs, cleaning, toilet preparations	1078	52	53	1183
30 Paints and allied products	566	170	177	912
31 Petroleum refining and related industries	5688	646	673	7006
32 Rubber & misc. plastic products	601	458	471	1530
33 Leather tanning & indust. leather prods.	40	18	18	76
34 Footwear and other leather products	15	13	13	41
35 Glass and glass products	146	81	81	309
36 Stone and clay products	561	1428	1504	3493
37 Primary iron and steel manu-facturing	3845	4479	4646	12,970
38 Primary nonferrous metal manufacturing	2980	2635	2744	8359
39 Metal containers	778	37	39	853
40 Heating, plumbing, struct. metal products	381	1367	1444	3193
41 Stampings, screw machine prods., bolts	282	398	403	1082
42 Other fabricated metal products	2327	5191	5478	12,995
43 Engines and turbines	223	198	207	628
44 Farm machinery and equipment	54	45	47	145
45 Construc., mining, oil field machinery	281	201	212	693
46 Materials handling mach. & equip.	147	131	138	416
47 Metalworking machinery & equip.	401	484	490	1375
48 Special indust. mach. and equip.	806	190	200	1196
49 General indust. mach. and equip.	9167	10,396	11,010	30,573
50 Machine shop products	184	796	512	1492
51 Office, computing, accounting machines	118	101	106	325

No.	Industry	Opera-tion	Replace-ment	Expan-sion	Total
52	Service industry machines	147	233	244	624
53	Electric industrial equip. & apparatus	631	2233	2360	5224
54	Household appliances	162	224	235	621
55	Electric lighting and wiring equip.	137	273	286	697
56	Radio, TV, communications equip.	122	151	157	431
57	Electronic components & accessories	101	145	148	393
58	Misc. electrical mach., equip., supplies	95	101	100	295
59	Motor vehicles and equipment	471	1185	879	2535
60	Aircraft and parts	149	162	168	479
61	Other transportation equipment	106	100	103	300
62	Scientific and controlling instruments	182	250	257	688
63	Optical, opthalmic, photog. equipment	66	30	31	128
64	Misc. manufacturing	250	117	121	488
65	Transportation and warehousing	4690	1661	1676	8027
66	Communications; exc. radio & TV	804	314	326	1444
67	Radio and TV broadcasting	239	94	98	432
68	Electric, gas, water, sanitary services	26,604	842	875	28,322
69	Wholesale and retail trade	4279	2299	2398	8977
70	Finance and insurance	1617	545	564	2726
71	Real estate and rental	2978	810	839	4626
72	Hotels, personal repair serv., exc. auto	285	157	138	580
73	Business services	3838	1514	1578	6929
74	Research and development	—	—	—	—
75	Auto repair and services	362	180	176	718
76	Amusements	134	52	54	241
77	Medic., educ., nonprofit orgs.	132	54	57	243
78	Federal gov't. enterprises	834	137	142	1114
79	State and local gov't. enterprises	3683	187	193	4062
	Total	174,571	61,074	63,303	298,945

Table 20. *Changes in Output Levels to Maintain
Improved Standards of Air and Water Pollution Abatement
in the Paper Industries, 1980
(thousands of dollars)*

No. Industry	Opera-tion	Replace-ment	Expan-sion	Total
1 Livestock & livestock products	1126	215	223	1564
2 Other agricultural products	1189	368	384	1940
3 Forestry and fishery products	354	191	201	747
4 Agric., forestry, fishery services	98	36	37	171
5 Iron & ferroalloy ores mining	965	419	435	1820
6 Nonferrous metal ores mining	1203	419	436	2058
7 Coal mining	2035	287	298	2619
8 Crude petroleum and natural gas	7904	626	652	9182
9 Stone & clay mining and quarrying	541	406	427	1374
10 Chemical & fertilizer mineral mining	3379	53	55	3487
11 New construction	—	19,145	20,260	39,405
12 Maintenance & repair construction	8151	580	597	9328
13 Ordinance and accessories	28	47	49	124
14 Food and kindred products	3235	442	459	4136
15 Tobacco manufactures	59	22	23	105
16 Br. & nar. fabrics, yarn, thread mills	349	219	221	789
17 Misc. textile goods and floor coverings	158	184	189	530
18 Apparel	167	96	99	363
19 Misc. fabricated textile products	153	49	44	246
20 Lumber & wood prod., exc. containers	1476	1940	2047	5463
21 Wooden containers	64	33	34	130
22 Household furniture	21	131	138	290
23 Other furniture and fixtures	12	70	74	157
24.01 Pulp	892	204	213	1310
24.02 Paper	837	211	220	1269
24.03 Paperboard	741	158	164	1063
24.04 Envelopes	66	20	20	106
24.05 Sanitary paper	36	18	19	73
24.06 Wall and building paper	72	49	52	173

No. Industry	Operation	Replacement	Expansion	Total
24.07 Converted paper n.e.c.	1241	211	221	1673
25 Paperboard containers and boxes	1142	332	345	1818
26 Printing and publishing	1851	695	723	3270
27 Chemicals and sel. chemical products	110,344	1223	1274	112,841
28 Plastics and synthetic materials	2901	372	385	3658
29 Drugs, cleaning, toilet preparations	1814	87	90	1991
30 Paints and allied products	952	286	297	1535
31 Petroleum refining & rel. industries	9570	1087	1133	11,789
32 Rubber and misc. plastic products	1011	772	792	2575
33 Leather tanning & indust. leather prods.	68	30	31	128
34 Footwear & other leather products	26	21	22	69
35 Glass and glass products	246	137	137	520
36 Stone and clay products	943	2404	2530	5877
37 Primary iron & steel manufacturing	6470	7539	7815	21,825
38 Primary nonferrous metal manufacturing	5014	4434	4617	14,065
39 Metal containers	1309	62	65	1436
40 Heating, plumbing, struct. metal products	642	2301	2430	5372
41 Stampings, screw machine prods., bolts	474	670	677	1822
42 Other fabricated metal products	3915	8734	9216	21,865
43 Engines and turbines	376	333	348	1057
44 Farm machinery and equipment	91	75	78	245
45 Construc., mining, oil field machinery	472	338	356	1167
46 Materials handling mach. & equip.	247	221	232	700
47 Metalworking machinery & equip.	674	814	825	2313
48 Special indust. mach. and equip.	1356	320	336	2012
49 General indust. mach. and equip.	15,425	17,493	18,506	51,424
50 Machine shop products	310	1348	866	2525
51 Office, computing, accounting machines	198	170	179	547

No. Industry	Opera-tion	Replace-ment	Expan-sion	Total
52 Service industry machines	247	393	411	1050
53 Electric industrial equip. & apparatus	1062	3759	3970	8792
54 Household appliances	273	376	396	1045
55 Elec. lighting and wiring equip.	231	460	482	1173
56 Radio, TV, communications equipment	205	255	265	725
57 Electronic components & accessories	169	244	249	662
58 Misc. electrical mach., equip., supplies	159	170	168	497
59 Motor vehicles and equipment	793	2004	1485	4283
60 Aircraft and parts	251	273	282	806
61 Other transportation equipment	178	168	174	519
62 Scientific & controlling equipment	305	420	432	1158
63 Optical, opthalmic, photog. equipment	112	51	52	215
64 Misc. manufacturing	421	196	204	822
65 Transportation and warehousing	7892	2798	2821	13,511
66 Communications; exc. radio & TV	1353	528	548	2429
67 Radio and TV broadcasting	403	159	165	727
68 Electr., gas, water, sanitary services	44,768	1418	1471	47,657
69 Wholesale and retail trade	7201	3870	4035	15,105
70 Finance and insurance	2721	917	949	4587
71 Real estate and rental	5011	1363	1411	7785
72 Hotels; personal repair serv., exc. auto	480	265	233	978
73 Business services	6458	2548	2654	11,660
74 Research and development	—	—	—	—
75 Auto repair and services	610	303	296	1208
76 Amusements	225	88	92	405
77 Medic., educ., nonprofit orgs.	223	92	95	410
78 Federal government enterprises	1404	231	239	1874
79 State and local gov't. enterprises	6197	315	324	6835
Total	293,742	102,808	106,502	503,049

Table 21. *Principal Industries Affected by Higher Abatement Standards in the Paper Industries (millions of dollars)*

Amount of change	1970		1980	
	Industry affected	Increase in output	Industry affected	Increase in output
more than 100			27 Chem. & sel. chem. products	113
50–99.9	27 Chem. & sel. chem. products	67	49 Gen'l. industrial mach. & equip.	51
25–49.9	49 Gen'l. industrial mach. & equip.	30	68 Elec., gas, water, sanitary services	48
	68 Elec., gas, water sanitary services	28	11 New construction	39
10–24.9	11 New construction	23	42 Other fabricated metal products	22
	37 Primary iron and steel manufacturing	13	37 Primary iron and steel manufacturing	21
	42 Other fabricated metal products	13	69 Wholesale and retail trade	15
			65 Transportation and warehousing	14
			38 Primary ferrous nonmetal manufacturing	14
			73 Business services	12
			31 Petroleum refining and related industries	11

With new levels in pollution abatement, the increases in outputs mentioned above result in corresponding increases in employment in each industry. Tables 22 and 23 list the estimated effect on employment. The total number of additional jobs predicted for 1980 as the result of pollution control in the paper-and-allied-products industry is 22,720. (The analogous figure for 1970 was 13,501 jobs.) Total national civilian employment is projected at 98,600,000 in 1980 [63]. Additional pollution control in the paper-and-allied-products industries would therefore contribute no more than approximately .04 percent to the national employment level.

5.2 Price Impact

The input-output price equation

$$R' = V'(I-A)^{-1} \tag{3}$$

in which R' is a row vector of price relatives
V' is a row vector of value added coefficients
A is the matrix of input coefficients
was solved to determine the price effect of the new abatement levels in the paper and allied products industry. The value-added coefficients per unit of pollutant abated included labor costs for operation of the additional pollution control equipment, plus a capital depreciation charge computed by the straight line depreciation method. Capital equipment life was estimated at ten years for particulate abatement equipment and 20 years for BOD_5 abatement equipment. The resulting price effect is nominal. The computed price increase of pulp is 2.3 percent; that of paper, 1.1 percent; that of paperboard, 1.7 percent; and that of construction board, 1.5 percent. Secondary price effects are virtually absent.

The absence of a price effect was a little surprising, especially in view of industry claims that more stringent abatement would require substantial price increases [22]. The figures quoted here are based on averages. For certain classes of paper the price effects may well be larger. Also for certain companies and certain plants that are operating below the industry average in abatement practices, the effect will be more severe. It may be difficult to fit abatement equipment to some older plants. In addition, only pollution abatement technologies with proven economy-

Table 22. *Employment Created by Increased Pollution Abatement in the Paper Industries, 1970 (number of man years)*

No. Industry	Opera-tion	Replace-ment	Expan-sion	Total
1 Livestock & livestock products	48	9	10	67
2 Other agricultural products	74	23	24	121
3 Forestry and fishery products	28	15	16	58
4 Agric., forestry, fishery services	11	4	4	20
5 Iron & ferroalloy ores mining	20	8	9	37
6 Nonferrous metal ores mining	31	11	11	53
7 Coal mining	63	9	9	81
8 Crude petroleum and nat'l. gas	43	3	4	50
9 Stone and clay mining & quarrying	21	15	16	52
10 Chemical & fertilizer mineral mining	61	1	1	63
11 New construction	—	565	597	1162
12 Maintenance & repair construc-tion	292	21	21	334
13 Ordinance and accessories	1	1	2	4
14 Food and kindred products	51	7	7	65
15 Tobacco manufactures	—	—	—	1
16 Br. & nar. fabrics, yarn, thread mills	9	6	6	21
17 Misc. textile goods and floor coverings	4	4	4	12
18 Apparel	8	4	5	17
19 Misc. fabricated textile products	6	2	2	9
20 Lumber & wood prod., exc. containers	61	80	84	225
21 Wooden containers	4	2	2	8
22 Household furniture	1	6	7	14
23 Other furniture and fixtures	1	3	3	6
24.01 Pulp	13	3	3	19
24.02 Paper	19	5	5	28
24.03 Paperboard	9	2	2	12
24.04 Envelopes	2	1	1	4
24.05 Sanitary paper	—	—	—	1
24.06 Wall and building paper	2	1	2	5

No. Industry	Opera-tion	Replace-ment	Expan-sion	Total
24.07 Converted paper n.e.c.	28	5	5	38
25 Paperboard containers and boxes	28	8	8	45
26 Printing and publishing	74	28	29	131
27 Chemicals and sel. chemical products	1835	20	21	1877
28 Plastics and synthetic materials	51	6	7	64
29 Drugs, cleaning, toilet preparations	27	1	1	30
30 Paints and allied products	14	4	5	23
31 Petroleum refining & rel. industries	53	6	6	66
32 Rubber & misc. plastic products	26	20	20	66
33 Leather tanning & industr. leather goods	2	1	1	3
34 Footwear and other leather products	2	1	1	4
35 Glass and glass products	8	5	5	18
36 Stone and clay products	28	73	76	178
37 Primary iron & steel manufac-turing	143	166	172	481
38 Primary nonferrous metal manufacturing	85	75	78	238
39 Metal containers	23	1	1	26
40 Heating, plumbing, struct. metal products	20	71	75	165
41 Stampings, screw machine prods., bolts	20	29	29	78
42 Other fabricated metal products	128	285	301	714
43 Engines and turbines	9	8	8	25
44 Farm machinery and equipment	3	2	2	7
45 Construc., mining, oil field machinery	13	9	10	32
46 Material handling mach. & equipment	7	6	7	20
47 Metalworking machinery and equipment	28	34	35	97
48 Special indust. mach. and equip.	50	12	12	74
49 General indust. mach. and equip.	493	559	592	1643
50 Machine shop products	18	78	50	147

No. Industry	Opera-tion	Replace-ment	Expan-sion	Total
51 Office, computing, accounting machines	6	5	5	15
52 Service industry machines	5	8	8	21
53 Electric industrial equip. and app.	36	127	135	298
54 Household appliances	5	7	7	20
55 Elec. lighting and wiring equip.	8	15	16	39
56 Radio, TV, communications equip.	6	7	7	20
57 Electronic components & accessories	6	9	9	25
58 Misc. electrical mach., equip., supplies	5	5	5	15
59 Motor vehicles and equipment	9	23	17	48
60 Aircraft and parts	8	9	9	26
61 Other transportation equipment	5	5	5	16
62 Scientific & controlling instruments	12	17	17	46
63 Optical, opthalmic, photog. equip.	4	2	2	7
64 Misc. manufacturing	16	7	8	31
65 Transportation and warehousing	299	106	107	511
66 Communications; exc. radio & TV	42	16	17	75
67 Radio and TV broadcasting	13	5	5	24
68 Electric, gas, water, sanitary services	615	19	20	655
69 Wholesale and retail trade	520	280	292	1091
70 Finance and insurance	132	44	46	222
71 Real estate and rental	29	8	8	45
72 Hotels, personal & repair serv., exc. auto	52	29	25	106
73 Business services	425	167	175	766
74 Research and development	—	—	—	—
75 Auto repair and services	34	17	16	67
76 Amusements	15	6	6	26
77 Medic., educ., nonprof. orgs.	21	9	9	39
78 Federal government enterprises	124	20	21	165
79 State and local gov't. enterprises	218	11	11	241
Total	6767	3310	3425	13,501

Table 23. *Employment Created by Increased Pollution Abatement in the Paper Industries, 1980 (number of man years)*

No. Industry	Operation	Replacement	Expansion	Total
1 Livestock & livestock products	81	15	16	113
2 Other agricultural products	125	38	40	203
3 Forestry and fishery products	46	25	26	98
4 Agric., forestry, fishery services	19	7	7	33
5 Iron & ferroalloy ores mining	33	14	15	62
6 Nonferrous metal ores mining	52	18	19	89
7 Coal mining	106	15	16	136
8 Crude petroleum and nat'l. gas	73	6	6	85
9 Stone and clay mining & quarrying	35	26	27	88
10 Chemical & fertilizer mineral mining	102	2	2	105
11 New construction	—	950	1005	1955
12 Maintenance & repair construction	492	35	36	563
13 Ordinance and accessories	2	2	3	7
14 Food and kindred products	85	12	12	109
15 Tobacco manufactures	1	—	—	1
16 Br. & nar. fabrics, yarn, thread mills	16	10	10	36
17 Misc. textile goods and floor coverings	6	7	7	20
18 Apparel	13	8	8	28
19 Misc. fabricated textile products	10	3	3	16
20 Lumber & wood prod., exc. containers	102	134	142	378
21 Wooden containers	7	3	3	13
22 Household furniture	2	10	11	23
23 Other furniture and fixtures	1	5	5	11
24.01 Pulp	22	5	5	32
24.02 Paper	32	8	8	48
24.03 Paperboard	14	3	3	21
24.04 Envelopes	4	1	1	6
24.05 Sanitary paper	1	—	—	1
24.06 Wall and building paper	4	2	3	9

No. Industry	Opera-tion	Replace-ment	Expan-sion	Total
24.07 Converted paper n.e.c.	47	8	8	63
25 Paperboard containers and boxes	47	14	14	75
26 Printing and publishing	124	47	49	220
27 Chemicals and sel. chemical products	3088	34	36	3158
28 Plastics and synthetic materials	85	11	11	107
29 Drugs, cleaning, toilet preparations	46	2	2	50
30 Paints and allied products	24	7	8	39
31 Petroleum refining & rel. industries	90	10	11	110
32 Rubber & misc. plastic products	44	33	34	111
33 Leather tanning & industr. leather goods	3	1	1	5
34 Footwear and other leather products	3	2	2	7
35 Glass and glass products	14	8	8	30
36 Stone and clay products	48	122	129	299
37 Primary iron & steel manufac-turing	240	280	290	810
38 Primary nonferrous metal manufacturing	143	127	132	401
39 Metal containers	39	2	2	43
40 Heating, plumbing, struct. metal products	33	119	126	278
41 Stampings, screw machine prods., bolts	34	48	49	132
42 Other fabricated metal products	215	480	506	1202
43 Engines and turbines	15	13	14	43
44 Farm machinery and equipment	4	4	4	12
45 Construction, mining, oil field machinery	22	16	16	54
46 Material handling mach. & equipment	12	11	11	34
47 Metalworking machinery and equipment	48	58	58	164
48 Special indust. mach. and equip.	84	20	21	124
49 General indust. mach. and equip.	829	940	994	2763
50 Machine shop products	31	133	85	249

No. Industry	Operation	Replacement	Expansion	Total
51 Office, computing, accounting machines	9	8	8	26
52 Service industry machines	8	13	14	36
53 Electric industrial equip. and apparatus	61	214	227	502
54 Household appliances	9	12	13	33
55 Elec. lighting and wiring equip.	13	25	27	65
56 Radio, TV, communications equip.	9	12	12	33
57 Electronic components & accessories	11	15	16	42
58 Misc. electrical mach., equip., supplies	8	9	9	26
59 Motor vehicles and equipment	15	38	28	81
60 Aircraft and parts	14	15	15	44
61 Other transportation equipment	9	9	9	27
62 Scientific & controlling instruments	21	28	29	78
63 Optical, opthalmic, photog. equipment	6	3	3	12
64 Misc. manufacturing	27	13	13	52
65 Transportation and warehousing	502	178	180	860
66 Communications, exc. radio & TV	70	27	28	126
67 Radio and TV broadcasting	22	9	9	40
68 Electric, gas, water, sanitary services	1036	33	34	1102
69 Wholesale and retail trade	875	470	490	1836
70 Finance and insurance	222	75	77	374
71 Real estate and rental	49	13	14	76
72 Hotels, personal & repair serv., exc. auto	88	49	43	179
73 Business services	714	282	294	1290
74 Research and development	—	—	—	—
75 Auto repair and services	57	28	27	112
76 Amusements	25	10	10	44
77 Medic., educ., nonprofit orgs.	35	15	15	65
78 Federal government enterprises	209	34	36	278
79 State and local gov't. enterprises	367	19	19	405
Total	11,386	5572	5762	22,720

wide applications and known cost factors are included here. Price effects of removing gaseous emissions would increase prices further.

5.3 Pollution Emissions

Table 24 lists emissions by the paper-and-allied-products industries in 1963, 1970, and 1980 on the basis of 1963 pollution abatement practices. Only the first three measures of pollution emissions—BOD_5, suspended solids, and ambient particles—are assumed to be abated. With the new abatement practices (BOD_5 abated 85 percent; ambient particles, 99.9 percent), the emissions are reduced to the levels shown in the bottom half of Table 24. Although these levels represent substantial improvements, it is evident that by 1980 additional improvements in abatement techniques must be available if the environment is to be kept clean. The anticipated 1980 residual levels of BOD_5 and suspended solids emissions come very close to the actual 1963 levels. (As indicated earlier, gaseous emissions have not been abated in the model.)

5.4 Conclusions

Recycling would mean a cleaner environment at a lower cost. When the reduction in solid waste is taken into account, an environment in which a maximum amount of paper and paperboard is recycled costs less to keep clean—even though some of the recycling activities (e.g., de-inking) are more water polluting than virgin fiber pulp manufacturing. Paper recycling would reduce emissions of large quantities of gases, most of which are both noxious and harmful (H_2S, RSH, RSR, SO_2, CO), and all of which are by-products of the most widely used pulping process.

On the other hand, virgin fiber pulp manufacturing is more

Table 24. *Emissions of the Paper and Allied Products Industries with 1963 and Proposed Abatement Policies*

		1963	1970	1980
With 1963 Abatement Policies				
BOD_5	(10^3 tons)	2154	2719	4575
Suspended solids	(10^3 tons)	1957	2462	4165
Gallons effluent	(10^9)	1742	2181	3703
Ambient particles	(10^3 tons)	190	238	404
Hydrogen sulphide	(10^3 tons)	13	16	28
Methyl mercaptan	(10^3 tons)	45	56	95
Dimethyl sulphide	(10^3 tons)	26	33	56
Sulphur dioxide	(10^3 tons)	30	38	64
Carbon monoxide	(10^3 tons)	360	451	764
With 85 percent Abatement of BOD_5 and 99.9 percent abatement of ambient particles				
BOD_5*	(10^3 tons)	451	569	957
Suspended solids*	(10^3 tons)	734	923	1562
Gallons effluent	(10^9)	1742	2462	4165
Ambient particles*	(10^3 tons)	2	2	4
Hydrogen sulphide	(10^3 tons)	13	16	28
Methyl mercaptan	(10^3 tons)	45	56	95
Dimethyl sulphide	(10^3 tons)	26	33	56
Sulphur dioxide	(10^3 tons)	30	38	64
Carbon monoxide	(10^3 tons)	360	451	764

*Pollutants whose abatement is greater under the new policies.

stimulating for the economy than paper recycling, i.e. paper recycling is cheaper. Directly and indirectly, the pulp mills have a greater need for goods and services from the economy than the paper recycling industries. Although these results are in part due to the nature of the data that were used in this study, the sign of the difference is correct even if the magnitude is overstated. Less employment is generated by recycling than by virgin fiber pulp manufacturing.

Although this study does not include an analysis of regional impact in the strict sense, it is evident that most of the positive impact of an increase in recycling—solid waste reduction and creation of simple-skill jobs—takes place in urban areas. Most of the undesirable impact of an increase in recycling affects the rural areas through less rapid expansion of pulp mills. Finally, of course, paper is a product that is most heavily consumed in urban areas. A national policy to emphasize recycling of waste-paper by imposing more stringent GSA specifications for recycle content should take the possible effects into account.

In this study, the impact on the U.S. economy of higher levels of pollution abatement practices in the paper and allied products industry has been assessed. The major direct and indirect supplier industries for new levels of abatement of BOD_5 and ambient particles have been identified. The largest absolute expansion (once the initial investment has been made) is required from the chemicals and selected chemical products industry; it amounts to $67 million in 1970 and $113 million in 1980. The increased employment necessary for new abatement practices seems nominal at .04 percent of the nation's civilian employment. In addition, the price increases that will result from the new levels of abatement appear small. In other words, higher levels of pollution abatement in the paper and allied products industry do not require any extraordinary efforts from the U.S. economy. It is evident that additional technologies of abatement must be developed relatively soon. Even with the new levels of abatement anticipated, 1980 BOD_5 discharge levels will reach close to 50 percent of the 1963 discharge level. In addition, an effective technology for the abatement of gaseous pollutants is badly needed.

REFERENCES

1. Abrahams, John H., "Wealth Out of Waste," *Nation's Cities* (September 1969).

2. Allan, Leslie et al., *Paper Profits* (Washington, D.C., The Council of Economic Priorities, 1971).

3. Almon, Clopper, "How to Purify an Input-Output Matrix," unpublished paper, 1967.

4. American Paper Institute, *Background Information on Recycling Waste Paper* (New York, 1971).

5. —— "Paper and Paper Manufacture," n.d.

6. —— *Paper, Paperboard and Woodpulp Capacity, 1970–1973* (New York, November 1971).

7. —— *The Paper Industry's Part in Protecting the Environment* (New York, n.d.).

8. —— unpublished tables and graphs (New York, July 1971).

9. (Joseph E.) Atchison Consultants, Inc., "A Preliminary Study of Waste Paper and Prospects for Its Increased Recycling," unpublished report prepared for the U.S. Department of Agriculture (November 1970).

10. Atchison, D. J. et al., "Future Prospects for Increased Waste Paper Recycling," *Paper Trade Journal* (September 13, 1971), pp. 58–69.

11. Bellows, Nancy, and John Goff, "Recycling Feasibility Study for the Municipal and Environment Services Component of Cambridge Model Cities" (Boston Environment, Inc., Boston, November 1970).

12. Billings, R. M., "The Other Side of the Recycling Coin," paper presented at the Recycling Seminar of the Institute of Paper Chemistry, Appleton, Wisconsin (November 1971).

13. Britt, Kenneth W., *Handbook of Pulp and Paper Technology* (New York, Reinhold Publishing Company, 1965).

14. Casey, James P., *Pulp and Paper: Chemistry and Chemical Technology*, 1–3 (New York, Interscience Publishers, 1961).

15. Cunningham, Roger H., "Secondary Fiber for Paperboard—How the Waste Paper Is Processed," *Paper Trade Journal* (September 13, 1971), pp. 65–67.

16. Crysler, Frederick S., "Report to the Board of Directors, American Paper Institute on the Solid Waste Council" [sic], Colorado Springs, Colorado (September 1971).

17. —— "The Demand for Recycled Fibers," *Tappi*, 54 (1971), 904–909.

18. Danforth, Donald W., "Refining—How to Gain Optimal Performance from Secondary Fiber," *Paper Trade Journal* (September 13, 1971), pp. 76–78.

19. Development Sciences, Inc., "A Technique for the Systematic Identification of Pollution Reduction Measures: EMIS," paper prepared for the U.S. Department of Health, Education and Welfare (1970).

20. Duprey, R. L., *Compilation of Air Pollutant Emission Factors*, U.S. Department of Health, Education and Welfare, U.S. Public Health

Service Publication No. GGG-AP-42 (Washington, Government Printing Office, 1968).

21. Edwards, Rodney J., "What's Ahead in Waste Paper Utilization," paper presented at the General Meeting of the Pulp and Raw Materials Group, American Paper Institute (March 1971).

22. Environmental Protection Agency, *The Economic Impact of Pollution Control* (Washington, Government Printing Office, 1972).

23. Environmental Protection Agency, New York, "Memorandum Regarding Packaging Legislation," undated.

24. Ernst and Ernst, "Costs and Economic Impacts of Air Pollution Controls," paper prepared for the U.S. Department of Health, Education and Welfare (October 1969).

25. Evans, John C. W., "What about Municipal Trash as a Source of Paper-making Fibers?" *Paper Trade Journal* (September 13, 1971), p. 79.

26. Felton, A. J., "Handling the Problem Papers—Wet Strength, Asphalt, Poly-coated, etc.," *Paper Trade Journal* (September 13, 1971), pp. 70–72.

27. Franklin, William E., and Arsen Darnay, *The Role of Nonpackaging Paper in Solid Waste Management, 1966 to 1976,* Environmental Protection Agency, Publication SW-26C (Washington, Government Printing Office, 1971).

28. Franz, Maurice, "More than Just a Disposal System," *Compost Science* (July–August 1970).

29. Gellman, Isaiah, "Fiber Reclamation Processes—What to Do with What's Left Over," *Paper Trade Journal* (September 13, 1971), pp. 73–75.

30. Gostenhofer, George, "Annual Report on Coefficient Projections," report to the Interagency Growth Project, Cambridge, Massachusetts, Harvard Economic Research Project (1968).

31. Graham, George A., "Inside the Waste Paper Market—Comments from a Dealer," *Paper Trade Journal* (September 13, 1971), pp. 68–69.

32. Hair, Dwight, "Use of Regression Equations for Projecting Trends in Demand for Paper and Board," U.S. Department of Agriculture, Forest Resource Report No. 18 (December 1967).

33. Hancock, William E., and Frank W. Lorey, "Testimony before the Committee on Rules and Administration of the Senate of the United States," unpublished paper (August 1971).

34. Harding, C. I., and J. E. Landry, "Future Trends in Air Pollution Control in the Kraft Pulping Industry," *Tappi,* 49 (1966), 61A–67A.

35. Hendrickson, E. R., et al., *Control of Atmospheric Emission in the Wood Pulping Industry,* 1–3, U.S. Department of Health, Education and Welfare, Reports PB 190351–53 (Springfield, Virginia, National Technical Information Service, March 1970).

36. Herbert, William, and Wesley A. Flower, "Waste Processing Complex

Emphasizes Recycling," *Public Works Magazine* (June 1971).

37. Hoch, Norman F., "Characteristics of Waste Papers Dictate Repulper Requirements," *Paper Trade Journal* (September 13, 1971), pp. 80–83.

38. Hodges, H. K. W., "The Situation Abroad—Great Britain," *Paper Trade Journal* (September 13, 1971), pp. 88–90.

39. Jacobs, Jane, *The Economy of Cities* (New York, Random House Publishing Company, 1970).

40. Kenline, Paul A., and Jeremy M. Hales, *Air Pollution and the Kraft Pulping Industry,* U.S. Department of Health, Education and Welfare, U.S. Public Health Service Publication 999-AP-4 (Washington, Government Printing Office, 1963).

41. Koenig, Franz, "The Situation Abroad—Germany," *Paper Trade Journal* (September 13, 1971), pp. 86–87.

42. Koenig, L. A., and P. M. Ritz, "Secondary Product Adjustment and Redistribution SPAR," NREC Technical Report No. 67 (Washington, D.C., National Planning Association, 1967).

43. Kruger, A., "Composition of the Gases Formed in the Sulfite Process," *Finnish Paper and Timber Journal,* Vol. 58, No. 2 (January 1935), 60–62.

44. Lannazzi, Fred, "Developments in Technology Influencing Secondary Fiber Use," *Paper Trade Journal* (April 17, 1972), pp. 42–44.

45. Lehto, Bjorn O., "Economic Parameters that Govern Secondary Fiber Usage, *Paper Trade Journal* (April 10, 1972), pp. 36–37.

46. Leonard, Edmund A., *Introduction to the Economics of Packaging* (New York, McGraw Hill Publishing Company, 1970).

47. Leontief, Wassily, "Environmental Repercussions and the Economic Structure: An Input-Output Approach," *Review of Economics and Statistics,* 52 (1970), 262–271.

48. MacLeod, Martin, "The Situation Abroad—Japan," *Paper Trade Journal* (September 13, 1971), pp. 84–85.

49. Marsh, Paul G., "Hydrasposal-Fibreclaim" (Middleton, Ohio, The Black Clawson Company, n.d.).

50. "Mills Can Make Recycling Work," *Paperboard Packaging* (February 1972), pp. 24–29.

51. Nutall, E. P., "Recycling Technology Linked to Economics," *Paperboard Packaging* (March 1972), pp. 44–46, 48, 65.

52. —— "Recycling—the Problem and the Progress," *Paperboard Packaging* (February 1972), pp. 32–36.

53. "Paper Fiber from Municipal Waste," *Paperboard Packaging* (February 1971), p. 19.

54. "Paper Makers Study Recycling," *Chemical Engineering* (December 14, 1970).

55. Perry, Henry J., "The Economics of Waste Paper Use," *Pulp and Paper* (April 1971), pp. 83–85.
56. —— "The Economics of Waste Paper Reuse: Part II," *Pulp and Paper* (May 1971), pp. 82–84.
57. "Rayonier Makes Major Breakthrough in Sulphite Pollution Control," *Paper Trade Journal* (May 15, 1972), pp. 50–51.
58. Stone, Richard, "Input-Output Relationships, 1954–1960," Paper No. 3, *A Programme for Growth* (Cambridge, University of Cambridge, 1963).
59. "Testing and Water Reuse Explored at Secondary Fibers Pulping Conference," *Tappi*, 52 (1969), pp. 2210–12.
60. U.S. Senate, *The Economics of Clean Air*, Report of the Administrator of the Environmental Protections Agency (Washington, Government Printing Office, 1971).
61. U.S. Department of Agriculture, *The Timber Owner and His Federal Income Tax*, Agriculture Handbook No. 274 (Washington, Government Printing Office, 1971).
62. U.S. Department of Commerce, *Input-Output Structure of the U.S. Economy: 1963*, Vols. 1–3 (Washington, Government Printing Office, 1969).
63. U.S. Department of Labor, *Patterns of U.S. Economic Growth*, Bulletin No. 1672 (Washington, Government Printing Office, 1970.
64. U.S. Department of the Census, *Census of Manufactures, 1963*, Vol. 2, *Industry Statistics* (Washington, Government Printing Office, 1966).
65. U.S. Department of the Interior, *The Cost of Clean Water*, Vol. 1, *Summary Report* (Washington, Government Printing Office, 1968).
66. —— *The Cost of Clean Water*, Vol. 2, *Detailed Analyses* (Washington, Government Printing Office, 1968).
67. —— *The Cost of Clean Water*, Vol. 3, *Industrial Waste*, Profile No. 3, "Paper Mills, Except Building" (Washington, Government Printing Office, 1968).
68. Vandegrift, A. E., et al., "Particulate Air Pollution in the United States," *Journal of the Air Pollution Control Association*, 21 (1971), 321–328.
69. Van Tassel, Alfred J., ed., *Environmental Side Effects of Rising Industrial Output* (Lexington, Massachusetts, Heath Lexington Books, 1970).
70. Weiner, Jack, *Air Pollution in the Pulp and Paper Industry* (Appleton, Wisconsin, Institute of Paper Chemistry, 1969).
71. "Will Industry Sell Recycle," *Modern Packaging* (September 1970), pp. 47–48.
72. Wilson, D. G., ed., "Summer Study on the Management of Solid Wastes," Cambridge, Massachusetts, Urban Systems Laboratory (Massachusetts Institute of Technology, September 1968).
73. Woodruff, M. D., "Centrifugal Cleaning of Secondary Fibers," *Tappi*, 54 (1971), 1148–51.

6

Steel Technologies, Pollution Abatement, and Local Air Quality

ELAINE BERLINSKY

1.0 INTRODUCTION

National input-output models [12, 30, 31] are useful for studying the general relationship between technological change,
national production, and atmospheric emissions, as well as the
cost implications of nationwide air pollution control. They tell
us nothing, however, about local effects. It is air pollution on
the local level that is of most concern to the public health of-
ficial and the private citizen. In addition to being a nuisance,
air pollution may damage health, impair visibility, destroy crops,
and cause erosion. Although some areas of the country are relatively free from air pollution, other areas are experiencing high
concentrations.

Local air pollution levels are related to topography, climate
and meteorology, population density, degree of urbanization,
industrial orientation, and technological as well as political
feasibility of air pollution control. Table 1 distinguishes between urban and nonurban concentrations of suspended particulates for the periods 1957–61 and 1962–66. One of the most
important factors contributing to local air pollution levels is the

Table 1. *Characteristics of Suspended Particulates: Urban and Nonurban*

Center-City Urban Characteristics of Suspended Particles

Site	1957–1961 Geom. Mean, mg/m³	1957–1961 Geom. Std. Dev., mg/m³	1962–1966 Geom. Mean, mg/m³	1962–1966 Geom. Std. Dev., mg/m³	Long-Term Trend	Change in Geometric Std. Dev.	Seasonal Pattern
Birmingham, Ala.	125.6	1.85	124.5	1.68	No change	Down[b]	Urban
Anchorage, Alaska	83.4	2.27	65.9	2.31	Down[b]	No change	Nonurban
Phoenix, Ariz.	206.5	1.58	165.7	1.73	Down[c]	No change	Urban
Little Rock, Ark.	74.1	1.63	87.0	1.74	Up[b]	Up[c]	None
Los Angeles, Calif.	164.2	1.56	124.5	1.60	Down[c]	No change	Urban
San Diego, Calif.	85.4	1.54	76.3	1.53	Down[b]	No change	Urban
San Francisco, Calif.	65.3	1.69	60.0	1.60	No change	No change	Urban
Denver, Colo.	137.4	1.64	125.1	1.54	No change	No change	Urban
Hartford, Conn.	88.8	1.59	95.8	1.58	No change	No change	Possible
New Haven, Conn.	84.8	1.47	91.9	1.52	Down[c]	Down[c]	Urban
Wilmington, Del.	175.3	1.64	131.1	1.41	Down[c]	Down[c]	None
Washington, D.C.	106.8	1.56	87.2	1.45	Down[c]	Down[c]	None
Tampa, Fla.	85.9	1.39	84.1	1.44	No change	No change	None
Atlanta, Ga.	99.4	1.58	93.4	1.49	No change	No change	None
Honolulu, Hawaii	48.8	1.50	38.7	1.37	Down[c]	No change	Urban
Boise, Idaho	104.7	1.53	80.3	1.53	Down[c]	No change	Urban
Chicago, Ill.	179.4	1.39	130.4	1.47	Down[c]	Up[b]	None
East Chicago, Ind.	176.7	1.70	183.7	1.51	No change	Down[b]	None
Indianapolis, Ind.	157.1	1.35	148.8	1.41	No change	No change	None

City						
Des Moines, Iowa	150.2	1.58	116.8	1.58	Down[c]	No change	Unusual
Wichita, Kan.	86.3	1.59	88.8	1.58	No change	No change	None
New Orleans, La.	88.4	1.37	85.2	1.44	No change	Up[b]	None
Portland, Me.	86.3	1.55	70.7	1.57	Down[c]	No change	None
Baltimore, Md.	131.5	1.51	130.3	1.49	No change	No change	Urban
Boston, Mass.	131.3	1.45	125.3	1.46	No change	No change	Urban
Detroit, Mich.	134.1	1.51	135.3	1.63	No change	Up[b]	None
Minneapolis, Minn.	94.4	1.75	74.8	1.50	Down[c]	Down[c]	None
Jackson, Miss.	71.7	1.60	69.1	1.46	No change	Down[b]	None
Kansas City, Mo.	140.7	1.50	129.3	1.48	No change	No change	Possible
St. Louis, Mo.	159.7	1.58	131.1	1.44	Down[c]	Down[c]	None
Helena, Mont.	54.7	2.01	48.7	1.73	No change	Down[c]	Unusual
Omaha, Nebr.	106.1	1.62	107.3	1.49	No change	Down[b]	Unusual
Newark, N.J.	97.2	1.63	103.8	1.54	No change	No change	None
Albuquerque, N.M.	183.5	1.71	114.6	1.70	Down[c]	No change	Possible
New York City, N.Y.	167.9	1.48	164.9	1.56	No change	No change	None
Charlotte, N.C.	114.4	1.59	101.3	1.62	Down[b]	No change	Urban
Bismarck, N.D.	80.0	1.77	78.7	1.94	No change	Up[b]	Unusual
Cincinnati, Ohio	124.6	1.45	129.2	1.50	No change	No change	Urban
Cleveland, Ohio	154.5	1.51	119.4	1.53	Down[c]	No change	None
Columbus, Ohio	129.0	1.51	108.1	1.48	Down[c]	No change	Unusual
Dayton, Ohio	113.0	1.49	117.2	1.64	No change	Up[b]	None
Youngstown, Ohio	137.7	1.56	126.1	1.56	No change	No change	Unusual
Portland, Ore.	75.5	1.77	85.7	1.93	No change	Up[b]	Unusual
Philadelphia, Pa.	162.3	1.52	155.5	1.41	No change	Down[b]	None
Pittsburgh, Pa.	160.3	1.73	150.6	1.55	No change	Down[c]	None
Providence, R.I.	100.1	1.54	106.9	1.48	No change	No change	None

Site	1957–1961 Geom. Mean, mg/m³	1957–1961 Geom. Std. Dev., mg/m³	1962–1966 Geom. Mean, mg/m³	1962–1966 Geom. Std. Dev., mg/m³	Long-Term Trend	Change in Geometric Std. Dev.	Seasonal Pattern
Columbia, S.C.	106.8	1.41	73.2	1.50	Down[c]	No change	None
Sioux Falls, S.D.	81.3	1.75	64.6	1.64	Down[c]	Down[b]	None
Chattanooga, Tenn.	190.1	1.55	154.4	1.50	Down[c]	No change	None
Nashville, Tenn.	126.6	1.57	116.3	1.57	No change	No change	Possible
Dallas, Tex.	91.4	1.71	91.6	1.58	No change	No change	Possible
Houston, Tex.	104.8	1.64	94.3	1.47	Down[b]	Down[c]	None
San Antonio, Tex.	105.9	1.71	72.0	1.52	Down[c]	Down[c]	Urban
Salt Lake City, Utah	105.6	1.64	108.3	1.63	No change	No change	Urban
Burlington, Vt.	50.6	1.53	56.8	1.61	Up[b]	No change	Nonurban
Norfolk, Va.	95.9	1.49	96.7	1.52	No change	No change	None
Seattle, Wash.	79.4	1.62	68.2	1.51	Down[c]	Down[b]	None
Charleston, W. Va.	171.2	2.20	169.0	2.07	No change	No change	Urban
Milwaukee, Wis.	139.4	1.49	120.0	1.63	Down[c]	Up[b]	None
Cheyenne, Wyo.	42.0	1.69	33.7	1.72	Down[c]	No change	Nonurban

Nonurban Characteristics of Suspended Particles

Site	1957–1961 Geom. Mean, mg/m³	1957–1961 Geom. Std. Dev., mg/m³	1962–1966 Geom. Mean, mg/m³	1962–1966 Geom. Std. Dev., mg/m³	Long-Term Trend	Change in Geometric Std. Dev.	Seasonal Pattern
Grand Canyon Co., Ariz.	18.7	2.42	18.5	2.11	No change	Down[b]	Nonurban
Montezuma Co., Colo.	11.8	2.15	12.2	2.60	No change	Up[b]	Nonurban
Kent Co., Del.	58.1	1.44	58.3	1.54	No change	No change	Nonurban
Butte Co., Idaho	18.6	2.08	13.8	1.91	Down[c]	No change	Nonurban
Parke Co., Ind.	51.3	1.43	49.3	1.58	No change	Up[b]	None

Location							
Delaware Co., Iowa	38.1	2.02	38.3	1.72	No change	Down^c	None
Acadia Nat'l. Park, Me.	24.1	1.83	22.3	1.83	No change	No change	Possible Nonurban
Calvert Co., Md.	37.8	1.83	39.1	1.48	No change	Down^b	Possible Nonurban
Glacier Nat'l. Park, Mont.	11.0	2.63	13.7	2.50	Up^b	No change	Nonurban
Thomas Co., Nebr.	22.3	1.89	19.4	2.04	No change	No change	Nonurban
White Pine Co., Nev.	10.9	2.61	10.0	2.59	No change	No change	Possible Nonurban
Coos Co., N.H.	18.3	1.61	19.6	1.77	Up^b	Up^b	None
Cape Hatteras, N.C.	31.5	1.41	48.4	1.81	Up^c	Up^c	None Nonurban
Ward Co., N.D.	20.0	2.22	31.8	2.19	Up^c	No change	None
Cherokee Co., Okla.	38.0	1.64	45.4	1.63	Up^b	No change	None Nonurban
Clarion Co., Pa.	36.6	1.67	37.1	1.69	No change	No change	None Nonurban
Washington Co., R.I.	30.2	2.07	37.4	1.99	Up^b	No change	Nonurban
Richland Co., S.C.	31.0	1.61	32.9	1.58	No change	No change	None
Orange Co., Vt.	38.5	1.43	38.7	1.59	No change	Up^b	Possible Nonurban
Shenandoah Nat'l. Park, Va.	29.8	1.53	30.1	1.59	No change	No change	Nonurban

^{b}Statistically significant.

^{c}Highly significant statistically.

Source: *Characteristics of Particulate Patterns 1957-1966*, Tables 1, 2 [60].

Table 2. *Global and U.S. Particulate Emissions*
(U.S. and Global Industrial Particulate Emissions, 1968)
(10^6 metric tons)

Industry	1968 Global Production[a]	U.S. 1968 % Global Production[b]	U.S. 1968 Uncontrolled Emissions[c]
Iron and steel gray iron foundries	915[h]	22	12.0
Grain handling, storage flour, feed milling	112[i]	10	2.3
Cement	513	13	7.9
Pulp and paper	90[j]	38	3.6
Miscellaneous			
Sand, stone, rock, etc.	—	—	2.4
Asphalt batching	—	—	2.5
Lime	—	—	0.8
Phosphate	—	—	0.4
Coal cleaning	—	—	0.5
Other minerals	—	—	0.4
Oil refineries	—	—	0.2
Other chemical inds.	—	—	0.3
Other primary and secondary metals	—	—	0.5
Total			33.8

[a] All data in this column from United Nations, *Statistical Yearbook*, 1969.
[b] Derived from United Nations, *Statistical Yearbook*, 1969, by taking the ratio of U.S. production to world production.
[c] NAPCA Division of Air Quality and Emissions Data, unpublished data.
[d] NAPCA, 1970.
[e] (Col. 3–Col. 4) ÷ Col. 3.
[f] Col. 3 [(100–Col. 2) ÷ Col. 2].
[g] Col. 3 + Col. 6.
[h] Combination of figures for production of pig iron and crude steel.

U.S. 1968 Controlled Emissions[d]	U.S. % Emissions Controlled[e]	Foreign 1968 Uncontrolled Emissions[f]	Global 1968 Uncontrolled Emissions[g]
1.890	84	42.6	54.6
1.890	84	42.6	54.6
1.020	56	20.7	23.0
0.790	90	53.0	60.9
0.660	82	5.9	9.5
	69		
0.790	—	—	—
0.490	—	—	—
0.410	—	—	—
0.185	—	—	—
0.170	—	—	—
0.160	—	—	—
0.090	—	—	—
0.085	—	—	—
0.085	—	—	—
6.825	81		

[i]Includes only production of wheat flour.
[j]Combination of figures for chemical and mechanical production of wood pulp.
Source: Study of Critical Environmental Problems, *Man's Impact on the Global Environment*, Tables 5.7, 5.8 [43].

area's industrial orientation. In 1967, 70 percent of the nation's steel was produced in 30 counties located in eight states. The iron and steel sector was responsible for 1.4 million tons of particulates, or nearly 10 percent of the particulates emitted from all industrial sources in the United States [37, 43] (Table 2).

This study uses an input-output model of the economy to demonstrate the relationship between air pollution controls in the steel industry, national production, and local air quality. The model has three components. The first is a national input-output table augmented to specify (1) air pollution abatement activities; (2) individual iron- and steel-making processes that are the predominant cause of particulate emissions in the integrated iron and steel industry; and (3) levels of abated and un-abated particulate emissions from these operations. The second is an algorithm for converting national emissions to local emissions. The third is a meteorological model that translates specific types of particulate emissions into local air quality.

Given this three-component model, along with data for the Gross National Product of goods and services, local production mix, and local meteorology, one can relate the gross national product of goods and services to local air quality (micrograms of particulates per cubic meter). Alternatively, standards of air quality can be set, and implications can be drawn about required levels of abatement as well as the inputs required to support them. The model can also be used to study price effects of various air pollution control strategies. This study will illustrate the diverse applications of the model and present an in-depth analysis of the data generated for iron and steel.

1.1 Processes and Resulting Emissions

Particulate air pollution in the integrated iron and steel industry is primarily caused by coke, iron, and raw steel production. The production technologies are described briefly below and the characteristics of the emissions resulting from these processes are presented in Table 3.

Table 3. *Nature of Emissions in Coke, Iron, and Raw Steel Production*

Process	Types of Emissions	When Do They Occur?	Abatement Measures	Comments
By-product coke plants	1) Settleable particulates (coal, coke, dust) 2) H_2S, SO_2 3) Phenols	1) Coal unloading and transfer 2) Charging, carbonization, and discharging 3) Storage and handling 4) Water quenching of incandescent coal	1) Remove H_2S 2) Oven and door maintenance 3) Baffle quench towers 4) Regulate coking times 5) Train employees in good housekeeping	None of the abatement measures is satisfactory. Future measures will include: 1) Negative oven pressure during charge 2) Larry car scrubbed 3) Hoods 4) Sealed larry cars 5) Coal through pipeline 6) Continuous coking 7) By-passing the process
Blast furnaces	1) Mostly iron oxide particulates 2) Other particulates are oxides related to the quality of the ore 3) CO	1) Slips, when pressure increases abnormally 2) Blast furnace gas 3) Handling	1) Prevent abnormal slips by periodically conducting controlled slips, by regulating rate of entry of charge 2) Decrease emissions during handling by hood arrangements and chemical additives	Pollution from blast furnace gas is minimal as the gas is collected, cleaned, and used as fuel.

Process	Types of Emissions	When Do They Occur?	Abatement Measures	Comments
Open hearth furnace	metallic oxides (type and concentration varies with time of cycle and nature of charge)	1) Melting 2) Tapping 3) Refining	1) Venturi scrubber 2) Electrostatic precipitator	Most open hearth furnaces are oxygen lanced. Oxygen lancing causes high production and emission rates.
Basic Oxygen Process furnace	same	1) Melting 2) Tapping 3) Refining	same	Good control is an integral part of furnace design.
Electric arc furnace	same	1) Charging 2) Melting 3) Tapping 4) Refining	1) Venturi scrubber 2) Electrostatic precipitator 3) Baghouse filter	Most emissions occur during charging, when roof swings and scrap is fed in. Screw-feed method of charge would prevent emission.

Sources: Barnes [6], Bramer [8], Brandt [10], Calenza [11], Franklin [20], Jablin [28], Schueneman [37], Shannon [39].

1.1.1 Coke Production

Bituminous coal is baked in by-product ovens until it is porous. The resulting material, coke, is acceptable for use in the blast furnace. Unlike coal, it burns on the inside as well as the outside and does not fuse into a sticky mass, but retains its strength under the weight of iron ore and limestone charged with it in the blast furnace. Coke production begins with the preparation and blending of coals. The prepared blend, typically 85–90 percent in sizes less than one-eighth of an inch is transferred to storage silos or bunkers on top of a battery (series) of coke ovens. To charge a coke oven, a weighted amount of prepared coal is lowered from the bunker to the larry car, which transports the coal to the coke oven. The coal is then loaded into the coke oven through slots on top of the oven. Approximately 15–20 hours after the coking process begins, good blast furnace coke is produced. The doors on either end of the oven are then removed and the coke is pushed out to be quenched and finally loaded into the blast furnace [4, 6].

1.1.2 Iron Production

A charge of iron ore, coke, and limestone is fed into the top of a blast furnace, a large steel stacklike chamber lined with refractory materials. As the solid materials descend, they are blasted with preheated air. The six major reactions that occur in the blast furnace are: (1) the dehydration of free water; (2) calcination of lime; (3) combustion of coke; (4) reduction of iron oxides to iron; (5) fusion of reduced iron; (6) the refining of iron by removing silicon, sulfur, and phosphorus. Every few hours the blast furnace is tapped from the bottom. The molten iron, which differs from steel only in its higher carbon content, is carried away from the furnace in hot-metal cars, while its impurities are discharged into a fluid slag [1].

1.1.3 Steel Production

Iron produced in the blast furnace is converted to steel in refining furnaces. Today, the bulk of the steel is produced in open hearth (OH) and basic oxygen process (BOP) furnaces, with the remainder in electric arc (EA) furnaces [2].

The basic oxygen process furnace used in the United States is barrel-shaped and tilts to receive a charge of molten blast furnace iron and cold steel scrap. After the furnace is turned upright, a cooled lance is lowered near the charged surface, and lime and other fluxes are added. Then high purity oxygen is blown through the lance at high pressure for twenty minutes. After the lance is withdrawn, the furnace is tilted to pour the molten scrap. The whole process takes about forty-five minutes— about one ninth as long as an open hearth "heat" (described below). The basic oxygen process furnace is rapidly replacing the open hearth as the workhorse of the industry [3, 42, 64].

The open hearth furnace is so called because the hearth or floor of the furnace, which is shaped like an elongated saucer, is exposed to a sweep of flames that melts the steel. To begin a heat, the furnace is charged with scrap, iron ore, and limestone. Flames, produced by oxygen gas combined with natural gas or liquid fuel, heat the charge for about two hours. Molten blast-furnace iron and cold scrap are then poured in and cooked six to ten hours [4, 69].

No molten blast-furnace iron is needed to produce steel in the electric arc furnace. Cold scrap is sorted out into as many as sixty-five types, and is loaded into the furnace. Huge electrodes are then lowered into the furnace. The electric current shoots boltlike arcs that melt the scrap, while gaseous oxygen and/or iron ore burn out the impurities. The whole furnace is tilted and the molten steel is poured out, with the entire process lasting four to six hours [4, 64].

2.0 GENERAL METHODOLOGY

The purpose of this study is to show how price and output levels in the national economy are related to various air pollution control strategies and local air pollution levels caused by coke, iron, and steel production activity. Figure 1 illustrates how the national economy, local emissions, and local air quality are tied together. Section 2.1 of this report describes component A of Figure 1; sections 2.2 and 2.3 describe components B and C, respectively.

2.1 The 102-order Input-Output Table: Component A

Three steps were taken to adapt the published input-output matrix to the needs of this study. The entire matrix is depicted

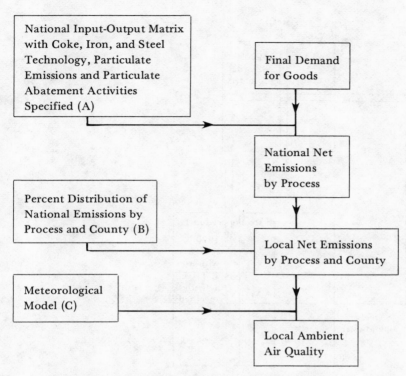

Figure 1. *General Scheme for this Study*

in Figure 2, block C. The first step was to expand the 83-order 1970 input-output table to 90 order by disaggregating input-output sector 37 to identify major activities in the primary iron and steel industry. Of greatest concern were those activities that are the predominant cause of particulate emissions to the atmosphere. Figure 2, section B, depicts this disaggregated 90-order matrix. The second step was to estimate emissions coefficients for the major particulate emitting processes and to incorporate these coefficients into Section A_{21} of the matrix. The third step was to specify air pollution abatement activities for emissions over the 1970 level. Costs per unit abated were entered into section A_{12}. Emissions from the air pollution abatement activities themselves, and thus all entries into the A_{22} portion, were set to zero.

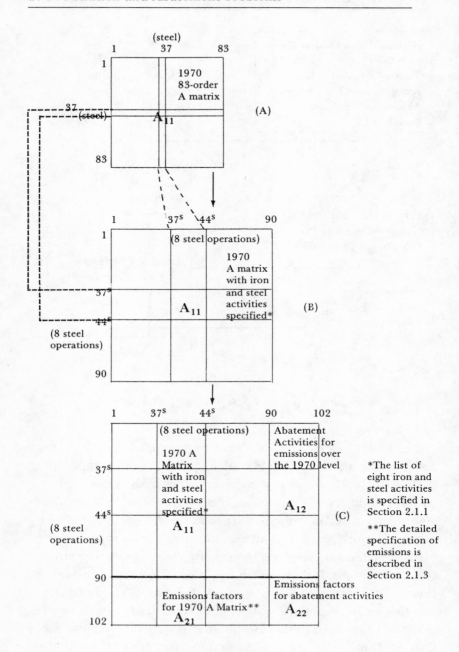

Figure 2. *The Augmented A Matrix*

2.1.1 The 90-order Technological Matrix A_{11}

The 1970 projected 83-order A matrix [55] was expanded to 90 order by disaggregating input-output sector 37 into eight subsectors. Seven of the subsectors denote operations within the integrated iron and steel industry in the United States. They consist of five subsectors representing the five operations that contribute to atmospheric particulate emissions (see 37^s–41^s, below). Together, these seven vectors correspond to input-output sector 37.01 of the 370-order input-output classification. The eighth subsector, 44^s, is comprised of iron and steel foundries, forgings, and miscellaneous operations. It corresponds to the sum of the 370-order input-output sectors 37.02, 37.03, and 37.04.

In summary, the disaggregation scheme for this study is:

83-order classification	370-order classification	90-order classification with steel disaggregated
		37^s coke production
		38^s blast furnace iron production
		39^s basic oxygen steel production
	37.01	40^s electric arc steel production
37		41^s open hearth steel production
		42^s in-house scrap production
		43^s rolling and finishing
	37.02, 37.03, 37.04	44^s iron and steel forgings, foundries, and misc. operations

This study is based on the projected 1970 83-order A matrix cited earlier. In order to derive 37^s–44^s, the more detailed 370-order iron and steel sectors were first substituted for input-output

Table 4. *Physical Inputs per Ton of Output of Coke, Iron, and Steel*

Input	Supplying Sector	Coke	Iron	Steel BOP	Steel EA	Steel OH	Input Units	Input Prices (1958 dollars)
Iron ore	5		1.568			0.035	tons	8.61
Coal	7	1.428					tons	5.64
Oxygen	27		0.095	1.890		0.964	10^6 cubic feet	2.39
Fuel	31	3.230	0.708			1.800	10^6 BTU	
Limestone	36		2.440	0.008	0.006	0.043	tons	1.05
Coke	37^S		0.626				tons	
Iron	38^S			0.848	0.026	0.606	tons	
Home scrap	42^S		0.024	0.214	0.638	0.305	tons	
Electricity	68			0.023	0.455	0.028	megawatt hours	9.47
Purchased scrap	83		0.015	0.131	0.390	0.187	tons	32.61

Sources for Table 4:

Column vectors for iron are directly from the American Iron and Steel Institute [2]. Coke and raw steel column vectors are directly from Tsao [47]. The following are exceptions:

(a) *Rows 42 and 90.* Approximately 35 percent of the scrap used is purchased, while the rest is home scrap [1]. Therefore:
home-scrap coefficients = (.65) x (Tsao's total scrap coefficients)
purchased-scrap coefficients = (.35) x (Tsao's total scrap coefficients)

(b) *Row 31, column "iron."* Tsao's data (15.44 MBTU per ton) on fuel input seemed to be too high for iron production when compared to real 1970 AISI data [2]. The AISI published input-output coefficients in terms of volumetric units of fuel per ton of iron produced in 1970. These were converted to millions of BTU, summed, and entered as fuel inputs.

fuel input	1970 AISI data (1)	Average heat content of fuel (2)	1970 AISI data converted (1) x (2)
natural gas	461 cubic feet per ton	$.1 \times 10^6$ BTU per cubic foot	$.461 \times 10^6$ BTU per ton
oil and tar	1.67 gallons per ton	$.148 \times 10^6$ BTU per gallon	$.247 \times 10^6$ BTU per ton
Total			$.708 \times 10^6$ BTU per ton

Prices assigned to physical coefficients were taken from the following sources:

(a) Prices of iron ore and limestone were taken directly from the 1958 *Minerals Yearbook* [61].
(b) Price of oxygen is derived from value and amount of pure oxygen shipments from SIC28 in the 1958 *Census of Manufactures,* Table 28A–17 [51].
(c) Price of electricity is directly form 1958 *Statistical Abstract of the U.S.* [53].
(d) Price of scrap is from Varga [69] in 1967 dollars ($26.30 per ton) and deflated to 1958 dollars using the ratio of 1967 to 1958 Wholesale Price Indices [58] for scrap (90.5:72.5).
(e) Price of fuel is based on average price per 10^6 BTU in 1958.

fuel	price (1958 dollars)	heat content	price per 10^6 BTU (1958 dollars)
natural gas	.119 per 10^3 cubic feet	10^3 cubic feet per 10^6 BTU	.119
No. 6 fuel oil	.109 per gallon	1 gallon per 10^6 BTU	.608
average			.364

sector 37. Then sectors 37.02, 37.03, and 37.04 were combined into 90-order sector 44^s, and 37.01 was disaggregated still further. A 370-order matrix is only available for 1963 [59], and we assume that input coefficients for sector 37.01 remain the same between 1963 and 1970.

Coefficients for the major components of column vectors 37^s to 42^s were determined by further disaggregation based on technical sources. Inputs per ton of output in physical units for each process are listed in Table 4 (columns 1–8, labeled $A_{i,37_s}$ through $a_{1,42_s}$). Coefficients of inputs from sectors 1 to 36, 43^s, 44^s, and 38 to 83 into sectors 37^s to 42^s were converted to dollars per ton of output; coefficients of sectors 37^s to 42^s into sectors 37^s to 42^s remained in physical units of "tons of input per ton of output." Inputs of sectors 1 to 36, 43^s, 44^s and 83^s into sector 43^s were calculated as residuals by subtracting the amounts purchased by 37^s to 42^s from the 370-order flows for 1970.*

Inputs of coke, iron, and scrap (sectors 37^s, 38^s, and 42^s) into rolling and finishing (sector 43^s) are assumed to be zero. Inputs of steel (39^s, 40^s, and 41^s) into 43^s were determined by dividing published production levels of basic oxygen process, electric arc, and open hearth steel by 1970 gross outputs for 370-order sector 37.01. Actual 1970 production levels, as well as predicted 1980 production levels of raw steel, are listed in Table 5. Column vectors 37^s to 43^s were also augmented to specify direct costs of air pollution control equipment already installed by 1970. A

* The 1970 technical coefficient for sales of chemicals to sector 37 [55] is about the same as the coefficient in the 1963 table [54]. Obviously it was not updated to account for additional oxygen consumed in basic oxygen process furnaces and in the oxygen-lancing of electric arc and open hearth steel. By multiplying physical coefficients (Table 4) by raw steel output (Table 5), we see that $442 million of oxygen was purchased in 1970 (1958 dollars). In 1963 purchases of oxygen accounted for $56 million, according to OBE Worksheets. In order to reflect this increase in oxygen consumption, inputs of sector 27 into 43^s were not calculated as the residual but were taken *in toto* from sector 37.01. According to the AISI [2], approximately 58 percent of fuel is consumed by sector 43^s, and 42 percent by sectors 37^s to 42^s.

Table 5. *Annual Raw Steel Production Data: 1970 and 1980*

Process	(1) Annual Prod. 1970 (millions of tons)	(2) Annual Prod. 1970, with 1980 Process Mix (millions of tons)	(3) Annual aver. Prod. 1980 Projection (millions of tons)	(4) Percent Distr. of Total Annual prod. 1970	(5) Percent Average Distribution of Total Annual Production 1980 Projection
BOP	63.33	84.81	107.25	47.7	64 (63–66)
EA	20.16	29.15	36.30	15.9	22 (20–24)
OH	48.02	18.55	21.45	36.3	14 (8–20)
Total	132.51	132.51	165.00 (154–175)	100	100

Sources:
column (1) American Iron and Steel Institute [2].
column (2) [column (3) x 132.51]/100.
column (3) Total steel projection (165 million tons is the average of two
projections: *33* magazine [68] predicts 154 million tons and
Hogan [23] predicts 175 million tons. Production by process =
column (5) x 165.00.
column (4) [column (5) x 132.51]/100.
column (5) Numbers in parentheses denote range predicted by *33* [68].

more detailed discussion of these adjustments will be found in
Section 2.1.3.

Row vectors for sectors 37^s through 43^s were determined by
assuming that all sales by the integrated iron and steel industry
to outside sectors come from sector 43^s, rolling and finishing.
Sales from rows 37^s through 42^s are set to zero. It is also assumed that the 370-order input-output structure remains the
same for the entire steel industry from 1963 to 1970. Therefore,
row 43^s is derived from the 1963 370-order row 37.01, and
row 44^s from the sum of rows 37.02, 37.03, and 37.04.

2.1.2 Particulate Emissions Coefficients (A_{21}): Data Collection

Particulate emissions from coke, iron, and steel production are estimated by multiplying production levels of these processes by predetermined emissions coefficients (in units of particulate emissions per unit of output). Unfortunately, published emissions coefficients are neither complete nor consistent. At best, one arrives at a rough estimate of uncontrolled emission rates, percent applicability and percent efficiency of control. These estimates are presented in Tables 6, 7, and 8.

Table 6 contains estimates of uncontrolled emissions rates. With respect to by-product coking, the environmental spokes-

Table 6. *Gross Emissions Rates*
(pounds of particulates per ton of output)

Source		Coke*	Iron	BOP	EA	OH
	[67]			45	30	17
Barnes	[4]		30–90	40	30	20–22
	[68]	14	180		15	17
Brandt	[7]					13
Calenza	[8]		200	40	10	9.3
	[70]	5				
USEPA	[57]		40–110	40	11	2
Oglesby	[30]			20–60		35
	[66]	14	80			
Vandegrift	[59]	3	130	40	40	17
mean		9	108	42	18	19
range		3–14	30–200	20–60	10–30	9–35

*Data for coking operations were collected in terms of "pounds of emissions per ton of coal consumed" and were converted to "pounds of emission per ton of coke produced." The conversion factor is 1.428 tons of coal per ton of coke.

Table 7. *Standard Efficiency of Air Cleaning Devices, by Process*

Source		Coke	Iron	BOP	EA	OH
				Process		
	[68]	0	98	99	98	95
Bramer	[6]	60	98	99+	98	98
Oglesby	[30]	0	98	99	98	95
	[67]	0	95	99+	99	97
Vandegrift	[50]	0	99	99+	99	97
mean		12	98	99+	97	96
range		0-60	95-99		95-98	92-98

man of a large steel corporation said:

> Insofar as we know, there is no confirmed emission factor
> for the coking process. The number of 2.0 lb/ton is a cal-
> culated emission factor and has been widely used and
> accepted in the literature. Because of the nature of the
> coking operation, actual measurements to determine the
> emission factor would require considerable research and
> very difficult measurement procedures. We know of no
> such effort to date, although some data may be generated
> as a result of the American Iron and Steel Institute – Office
> of Air Programs joint study on coke-oven-charging emission

Table 8. *Percent Application of Air Pollution Control Devices, 1970*

Source		Coke	Iron	BOP	EA	OH
				Process		
Bramer	[69]	5	100	100	75	50
Vandegrift	[59]	0	100	100	79	41
mean		3	100	100	77	46
range		0-5			75-79	41-50

control in Pittsburgh. We can neither confirm nor deny the figure of 10 lb/ton of coal which you quote, although we suspect it will turn out to be rather high. [5]

Blast furnace emissions are also difficult to estimate. Emissions occur from two sources: blast furnace slips and stack gases. The environmental spokesman of the same steel corporation commented: "With regard to blast furnace slips, here again [we] have no data on emission factors nor do we know of any. Our operating people state that some blast furnaces do not slip at all while others slip much more frequently than the figure you quote of 50 times per year. In general, we feel that 50 times per year is an acceptable number for calculating purposes and that it represents a well-run plant" [5]. Particulate emissions from blast furnace slips occur sporadically and last for less than two minutes. Although these emissions contribute greatly to temporary air pollution, they do not contribute significantly to annual levels of emissions. The major source of emissions, 30–200 pounds per ton of iron produced, is the stack gas.

Table 7 lists the percent standard efficiency of air cleaning in the iron and steel industry. Klauss [29] has estimated the cost of air cleaning devices with the following efficiencies of collection (by weight) for iron and steel production.

Process	Percent Efficiency
iron	99.99
BOP	99.40
EA	99.99
OH	92.00

Coefficients for controlled emissions (Section 2.1.3) and for abatement activities (section 2.1.4) are based on these four collection efficiencies.

Table 8 cites the percent application of air pollution control devices in 1970. Control devices have hardly been used in by-product coking. With increased oxygen lancing, the percentage of utilization of control devices has been increasing in open hearth and electric arc steelmaking. One large steel corporation reported in 1972 that 100 percent of its iron production operations, as well as 100 percent of its basic oxygen process and open hearth steel operations, were controlled. The steel corporation also reported that 100 percent of its electric arc furnaces were expected to be controlled by 1973. If this firm's air pollution abatement activities are typical of the entire industry's, it seems reasonable to predict that there will be 100 percent application of air pollution control devices on iron and steel furnaces by 1980, if not sooner.

2.1.3 Entry of Data into the Augmented Input-Output Matrix

Generally speaking, there are only two types of emissions for a given process: those collected, and those that escape into the atmosphere. Gross emissions are the sum of these two types. This model also distinguishes between two types of collected emissions for each process: those already collected by existing 1970 technology (with costs implicitly included in matrix A_{11}) and those that could have been collected by increased application of air pollution control (matrix A_{12}). For each process it also distinguishes two or three types of escaping particulate emissions with different physical characteristics. We make this distinction in order to calculate ambient air quality concentrations. For a detailed explanation of this calculation, see section 2.3.

In A_{21} (of the 102-order matrix), coefficients of particulate emissions from each air-polluting operation are specified in detail. (See Table 9.) There are 12 rows. Entries into row 91 denote average gross emissions before any controls are introduced. Entries into row 92 are average emissions already controlled by the industry in 1970. These latter entries correspond to the average gross emissions rates in Table 6, multiplied by corresponding factors in Tables 7 and 8. Because the percentage of control efficiency and control application for coking are negligible, controlled emissions from coking are set at zero.

Table 8 shows that there was already 100 percent application

Table 9. *Specifications of A_{21}*

	Coke 37^s	Iron 38^s	BOP 39^s	EA 40^s	OH 41^s
91 Gross	X	X	X	X	X
92 Already controlled (1970)	X	X	X	X	X
93 EA control over 1970		X	X	X	X
94 OH control over 1970					X
95 Coke uncontrolled	X				
96 EA uncontrolled				X	
97 OH uncontrolled					X
98 EA uncontrollable				X	
99 Iron wet scrubbers		X			
100 BOP wet scrubbers			X		
101 EA fabric filters				X	
102 OH electrostatic precipitators					X

of controls for iron and basic oxygen process steel emissions in 1970, whereas application of control was incomplete for coke, electric arc, and open hearth steel emissions. There is no present technology for controlling coke emissions, but technology does exist for removing most electric arc and all open hearth emissions. Entries into rows 93 and 94 denote emissions from electric arc and open hearth steel operations that were not controlled by 1970 but which could go on for further air cleaning. Thus, by definition, entries in rows 93 and 94 are zero for 1970.

Entries in rows 95 through 102 are net emissions, those that are uncontrolled and finally escape into the atmosphere. Such emissions from coke, electric arc, and open hearth operations appear in rows 95 through 97. The entry into row 98 represents uncontrollable emissions from electric arc operations (about 3 percent of gross emissions). The last four rows describe emissions resulting from the less-than-100-percent efficiency of the air cleaning devices in use.

2.1.4 Pollution Abatement Activities: A_{12}

2.1.4.1 Data Collection

Capital and operating costs of commonly used air-cleaning devices in the iron and steel industry were obtained from Klauss [29] and aligned with input-output 83-order classifications. (See Table 10.) Costs were taken for average-sized furnaces fitted with single, separate air-cleaning devices. Grain loadings after cleaning are given as .005 gscf for iron and .05 gscf for steel production. Estimates of grain loadings before cleaning were obtained from various sources. Efficiency of collection was then derived. Costs of cleanup per unit of emissions were calculated from equation (1) using data in Tables 6, 7, and 10:

$$\frac{\text{cost of cleanup}}{\text{tons emissions abated}} = \frac{\dfrac{\text{cost of cleanup}}{\text{tons output}}}{\dfrac{\text{tons emissions}}{\text{tons output}} \times \dfrac{\text{tons emissions abated}}{\text{tons emissions}}} \tag{1}$$

2.1.4.2 Entry of Data into the Augmented Input-Output Matrix

Pollution controls already in effect in 1970 were lumped with ordinary operating costs. Abatement above 1970 levels was assigned to special pollution abatement sectors. Table 11 shows the percentage distribution of installed pollution control devices (where air pollution control was being implemented) in 1970. Given these data, along with the 1970 percent application of control (Table 8) and the direct costs of cleanup per unit of output (Table 10), the actual expenditures on pollution control per

Table 10. *Costs of Air Cleaning, Using Wet Scrubbers (W.S.) and Electrostatic Precipitators (E.P.)*

1) in single average-sized iron and steel making furnaces (thousands of 1969 dollars)
2) per unit of output (1969 dollars per ton)
3) per unit of particulate emissions abated (1969 dollars per ton abated)

Input	Supplying Sector		Iron W.S.	OH E.P.	BOP W.S.	BOP E.P.	EA W.S.	EA E.P.	EA Fabric Filter
Capital Costs									
Foundation	11	1)	21.400	6.400	58.400	48.000	12.000	18.200	5.200
		2)	.046	.032	.027	.022	.040	.027	.017
		3)	.856	3.358	1.268	1.033	4.193	2.830	1.782
Structure	40	1)	50.000	28.800	87.600	96.000	21.000	16.400	18.200
		2)	.109	.144	.041	.045	.070	.055	.061
		3)	2.028	16.803	1.925	2.113	7.337	5.765	6.394
Water treatment	40	1)	178.400	28.800	452.600	112.000	91.100	13.700	5.200
		2)	.390	.144	.212	.053	.300	.046	.017
		3)	7.255	16.803	9.954	2.485	31.445	4.822	1.782
Duct and stack	40	1)	71.400	80.000	438.000	576.000	66.100	112.100	90.800
		2)	.156	.400	.205	.270	.220	.374	.303
		3)	2.904	46.676	9.626	12.668	23.060	39.202	31.760
Fan, motor and collector	49	1)	328.200	134.400	277.400	576.000	111.200	84.800	106.400
		2)	.718	.672	.130	.270	.371	.283	.354
		3)	13.357	78.412	6.014	12.668	38.888	29.664	37.106

Electrical	55	1)	42.800	32.000	102.200	128.000	30.100	27.300	23.400
		2)	.094	.160	.048	.060	.100	.091	.078
		3)	1.749	18.670	2.254	2.817	10.504	9.538	8.176
Control instruments	62	1)	21.400	9.600	43.800	64.000	9.000	10.900	10.400
		2)	.046	.048	.021	.030	.030	.036	.035
		3)	.856	5.601	.986	1.409	3.145	3.773	3.669
Central engineering	73	1)	360.000	110.000	390.000	410.000	95.300	88.800	82.400
		2)	.787	.550	.183	.192	.318	.300	.275
		3)	14.640	60.840	8.593	8.991	33.332	31.441	28.825
Client engineering	VA	1)	90.000	30.000	100.000	100.000	23.500	22.300	20.600
		2)	.197	.150	.097	.047	.078	.074	.069
		3)	3.665	17.502	2.208	2.206	8.176	7.757	7.232
Labor	VA	1)	936.000	170.000	790.000	800.000	163.000	147.600	122.900
		2)	2.048	.850	.370	.375	.543	.491	.410
		3)	38.099	99.182	17.373	17.608	56.916	51.466	42.967

Current Account Costs

Maintenance supplies	12	1)	20.000	10.000	44.000	46.400	9.400	8.500	9.200
		2)	.044	.050	.021	.022	.031	.028	.031
		3)	.818	5.834	.986	1.032	3.249	2.935	3.249
Electricity	68	1)	10.000	15.000	432.000	210.000	102.400	17.600	23.500
		2)	.022	.075	.203	.098	.341	.059	.078
		3)	.409	8.751	9.532	4.601	35.743	6.184	8.176
Maintenance labor	VA	1)	30.000	15.000	66.000	69.600	14.100	12.700	13.800
		2)	.066	.075	.031	.033	.047	.042	.046
		3)	1.228	8.751	1.456	1.550	4.926	4.299	4.822
Operating labor	VA	1)	20.000	30.000	60.000	30.000	35.300	17.600	17.700
		2)	.044	.150	.028	.014	.118	.059	.058
		3)	.818	17.502	1.298	.657	12.369	6.184	6.080

Table 11. *Percent Distribution of Pollution Control Devices by Type, 1970*

	Process		
Type of Device	BOP	EA	OH
Wet scrubber	0	11	0
Electrostatic precipitator	42	7	100
Baghouse filter	58	82	0

Source: Shannon [39].

unit of output for sectors 38^s–41^s were calculated and then deflated to 1958 dollars. These values were assumed to be originally included in sector 37.01. Thus they were removed from column 37.01 of the matrix and added to columns 38^s–41^s.

The unit costs of additional cleanup for electric arc and open hearth emissions not yet controlled in 1970 were entered in columns 93 and 94. These sectors can abate those emissions from electric arc and open hearth, respectively. In 1970 blast furnace iron and basic oxygen process steel were already controlled at highest feasible efficiency level with 100 percent application.

2.2 Local Emissions Levels: Component B

Since particulate emissions are a local phenomenon, it was necessary to relate the national emissions generated in the 102-order matrix described in section 2.1 to local emissions. The following simplifying assumptions were made:

— Local production capacity is a fixed percent of national production capacity.
— Local production levels are proportional to local production capacity.
— Local gross particulate emissions are proportional to local production levels.

— All localities have uniform percentage application and efficiency of air pollution control devices, equal to the national average.

The next two sections describe how the percentage of emissions of national particulates was estimated for the fifty-six countries where coke, iron, or steel was produced in the United States in 1970, and for the fifty-five counties where production is predicted for 1980. The present study relies heavily on exogenous information about industry's plans for expanding and retiring capacity in specific locations. Alternatively, a more general interregional model might have been used to predict local emissions.

2.2.1 Local Emissions Levels for 1970

Capacities for 1970 coke, iron, and steel production are published in the *Directory of the American Iron and Steel Institute* [3]. These capacities are listed primarily not by geographic areas but by company and name of installation. Considerable effort was required to translate the information in the directory into estimates of available capacity for each process in each county in the United States.

Local coke emissions in a certain area are assumed to be a function of the percentage of U.S. coke ovens in that area, the capacity of the coke ovens, and the national emissions level. Coke ovens, of course, do not all have the same capacity. Ovens may range from 12 to 22 feet in height, 40 to 50 feet in length, and 14 to 20 feet in width. There are no comprehensive data on the exact dimensions of coke ovens in the United States. Barnes [6] distinguishes, however, between conventional short ovens (12 feet high) and newer, tall ovens (at least 20 feet high). A search through the annual reports of *Iron and Steel Engineer* since 1960 reveals that no tall coke ovens were built before 1969. Inland Steel [27] reports that its new battery of 51 tall coke ovens is producing 40 percent more coke per day than an old battery of 51 low ovens. In this study, all tall coke ovens are

Table 12. *Estimated Distribution of Particulate Emissions by Process and by County, 1970 and 1980*

Percentage of National Emissions

State	County	Coke 1970	Coke 1980	BF 1970	BF 1980	BOP 1970	BOP 1980	EAF 1970	EAF 1980	BOH 1970	BOH 1980
Ala.	Etowah	1.0	1.0	0.8	0.8	2.1	1.7	2.2	1.8	4.7	—
Ala.	Jefferson	8.4	8.2	6.4	6.4	—	2.3	—	—	0.7	—
Calif.	Almeida	—	—	—	—	—	—	—	—	—	—
Calif.	Los Angeles	—	—	—	—	—	—	2.0	1.6	—	—
Calif.	San Bernardino	2.5	2.5	1.7	1.9	2.3	1.9	—	8.9	2.1	—
Colo.	Pueblo	1.6	1.5	1.4	1.7	1.7	1.4	—	1.5	1.2	—
Del.	New Castle	—	—	—	—	—	—	1.7	0.8	—	—
Ga.	Fulton	—	—	—	—	—	—	1.0	5.8	—	—
Ill.	Cook	2.4	2.3	7.5	7.5	6.2	6.5	4.1	1.9	5.4	—
Ill.	Madison	1.0	1.1	1.1	1.1	3.1	2.5	2.3	1.5	—	—
Ill.	Peoria	—	—	—	—	—	—	1.7	1.5	0.4	—
Ill.	Whiteside	—	—	—	—	—	—	4.7	6.0	—	—
Ind.	Howard	—	—	—	—	—	—	1.7	1.5	—	—
Ind.	Lake	13.1	12.9	9.3	9.3	7.9	11.5	1.4	1.2	15.9	22.4
Ind.	Porter	0.9	1.8	1.1	2.2	4.2	5.1	—	0.6	—	—
Ky.	Campbell	—	—	—	—	—	—	0.9	0.7	0.8	—
Ky.	Davis	—	—	—	—	—	—	0.7	0.6	—	—
Md.	Baltimore City	6.0	5.9	5.3	5.3	2.8	5.8	1.0	0.9	6.1	12.0
Mich.	Wayne	4.8	4.2	10.2	10.2	14.4	12.7	2.9	2.4	—	—
Minn.	St. Louis	0.9	0.9	0.6	0.6	—	—	—	—	1.6	—
Mo.	Jackson	—	—	—	—	—	—	3.2	2.7	—	—
N.J.	Burlington	—	—	—	—	—	—	0.7	0.6	—	—
N.Y.	Erie	5.4	5.0	5.3	5.3	6.2	5.5	1.1	—	3.6	—
N.Y.	Niagara	—	—	0.2	0.2	—	—	—	—	—	—

State	County										
N.Y.	Rensselaer	1.4	1.4	0.3	0.3	2.8	2.3	3.3	2.7		
Ohio	Butler	2.8	2.8	1.4	1.4	6.4	5.1	1.7	1.5	2.1	8.9
Ohio	Cuyahoga	2.5	2.5	5.0	5.0	4.2	3.4			1.8	
Ohio	Jefferson	1.9	1.8	2.1	2.1		1.5			2.7	
Ohio	Lorain	1.2	1.2	2.4	2.4					2.6	
Ohio	Lucas	3.8	3.0	0.7	0.7	2.1	4.5			1.8	
Ohio	Mahoning			1.9	1.9					6.1	
Ohio	Richland	0.5	0.5	1.1	1.1			1.2	1.0	1.1	
Ohio	Scioto	0.2	0.2	0.6	0.6	5.3	1.7	10.7	8.9		
Ohio	Stark	0.6		4.8	4.8	5.6	6.0	4.8	4.0		
Ohio	Trumbull	14.2	14.0	10.9	10.9			7.7	6.4	15.7	16.8
Pa.	Allegheny	2.1	2.1	3.8	3.8						
Pa.	Armstrong	3.1	3.7	2.2	2.2			5.7	4.7		
Pa.	Beaver	1.4	1.4	2.4	2.4				2.4	4.1	16.7
Pa.	Bucks	2.5	2.5					8.4	6.9		
Pa.	Cambria					10.2	8.2	2.0	1.7		
Pa.	Chester							2.7	2.2	1.9	
Pa.	Dauphin							1.1	0.9	1.0	
Pa.	Erie										
Pa.	Lawrence			6.8	0.8			1.3	1.1	0.2	
Pa.	Mercer					2.1	1.7	1.0	0.8	6.0	
Pa.	Mifflin										
Pa.	Montgomery	1.2	1.2	0.5	0.5	2.0	1.6	1.4	1.1	0.1	
Pa.	Northampton	3.0	2.9	2.5	2.5	3.5	2.9				
Pa.	Westmoreland	0.7	0.7	1.2	1.2						
S.C.	Georgetown							0.7	0.6		
Tenn.	Hamilton	0.8	0.3								
Texas	Harris	0.4	0.4					4.8	5.8	0.8	
Texas	Morris	0.6	0.6						6.0	1.4	
Utah	Utah	2.0	2.0	0.6	0.6					3.9	16.9
W.Va.	Marion	0.4	0.4	1.7	1.7						
W.Va.	Hancock	2.6	3.6	2.3	2.3	4.7	3.8			2.5	

thus assumed to be 40 percent more productive than lower ones. If a firm owns a battery of 100 tall coke ovens, they are entered into the table as a battery with the capacity of 140 conventional low ovens.

Local iron emissions depend on the percentage of working volume of all the blast furnaces in a certain area and the national emissions levels from iron production. Glover [22] reports that "in the 60's blast furnace men shifted from tons per square foot of open hearth to tons per 100 cubic feet of working volume as the yardstick for measuring [blast] furnace production."

Local steel emissions depend on the percent of all the basic oxygen process, open hearth, and electric arc furnace capacity (measured in tons per heat) in an area and the national emissions levels for each of these processes. Furnaces with capacities less than 100 tons per heat are omitted. All basic oxygen process furnaces are assumed to have an average heat time of 45 minutes; all open hearth furnaces, eight hours; and all electric arc furnaces, four hours [64].

Table 12 lists the estimated regional distribution of particulate emissions by process and by county for 1970 and 1980.

2.2.2 Local Emissions Levels in 1980

Exactly where new installations will be located and which old ones will be phased out is not easy to predict. A search of the literature, summarized in Table 13, showed that facilities are on-stream at various sites.

Coke Ovens. Although Bethlehem and Inland have announced that they will replace existing coke batteries at two installations, only two other firms report intentions of creating additional coking capacity [26, 27].

Blast Furnaces. Both *ISE* [26, 27] and *33* [68] report that an additional blast furnace will be erected at Bethlehem Steel in Porter county, Indiana. We assume that it will have the same capacity as one recently constructed at this site.

Steel Furnaces. To account for open hearth steel production

in 1980, *33* [68] reports that certain open hearths will remain
and that no new ones will be built. Although he fails to name the
installations, Tietig [44] reports that nine installations of open
hearths will probably remain in operation until 1980 and seven installations will probably remain until 1985. U.S. Steel at Cook,
Lorain, and Allegheny counties will be jointly responsible for 8
million tons of basic oxygen steel in 1985 [68] or approximately
2.6 million tons each, and National Steel at Wayne and Inland Steel
at Lake counties for 4 million tons. Annual capacity in terms of
tons per heat are estimated for these installations in Table 17.
We assume that basic oxygen process furnaces operate 8000 hours
per year at .75 hours per heat.

Bethlehem at Baltimore County is expected to replace 6600
tons per heat of open hearth capacity with basic oxygen process,
and U.S. Steel at Jefferson county will replace 4400 tons per
heat. Assuming that an open hearth heat takes eight hours and a
basic oxygen process heat .75 hours, the Bethlehem and U.S.
Steel installations will have about 620 and 410 ton-per-heat
capacities of basic oxygen process steel, respectively.

Electric arc furnaces will be installed in seven locations. The
capacities are only published [44, 46] for three of the seven installations. We assume that an open hearth heat takes eight hours
and an electric arc heat four hours, and that all electric arc
capacity will replace but not increase existing open hearth
capacity. Under these assumptions, CF&I's 1066 ton-per-heat
capacity of open hearth will be replaced by about 530 ton-per-
heat capacity of electric arc, and U.S. Steel at Dauphin will
replace its 777 ton-per-heat capacity of open hearth by about
390 of electric arc. It is extremely difficult if not impossible to
estimate the capacities of the electric arc furnaces to be built by
U.S. Steel in Cook and Bucks counties. Both installations are
scheduled to have two such furnaces in operation by 1980. The
Chicago plant presently has 3700 tons per heat—about 4 million
tons per year of open hearth capacity. The basic oxygen process
furnaces on stream for this plant will replace 2.6 million tons of
open hearth capacity. This would leave 1.4 million tons to be

Table 13. *Predicted New Installations in the Integrated Iron and Steel Industry, 1980*

Process	Source	Installation (State, County, Firm)	Effective Additional Capacity[a]
Coke	*ISE* [22]	West Virginia, Hancock, National	120 ovens
		Indiana, Porter, Bethlehem	115 ovens
		Total	235 ovens
Iron	*ISE, 33*	Indiana, Porter, Bethlehem	?
OH[b]	*33* [64]	Ohio, Butler, Armco	1860
		Pennsylvania, Allegheny, U.S. Steel	3520
		Pennsylvania, Bucks, U.S. Steel	3555
		Texas, Morris, Lone Star	1250
		Utah, Utah, U.S. Steel	3550
		Maryland, Baltimore, Bethlehem	2520
		Indiana, Lake, Youngstown	2400
		Indiana, Lake, Inland	2310
		Total	20965
BOP	*ISE, 33*	Illinois, Cook, U.S. Steel[d]	275
	[22, 64]	Ohio, Lorain, U.S. Steel[d]	275
		Pennsylvania, Allegheny, U.S. Steel[d]	275
		Michigan, Wayne, National[d]	200
		Indiana, Lake, Youngstown[d]	325
		New York, Erie, Republic[d]	100
		Indiana, Lake, Inland[e]	200
	ISE [21]	Indiana, Lake, U.S. Steel[f]	425
	ISE [22]	Maryland, Baltimore, Bethlehem[c]	620
		Indiana, Porter, Bethlehem[f]	325
		Alabama, Jefferson, U.S. Steel[c]	410
		Total	3430
EA[g]	*33* [64]	Texas, Harris, U.S. Steel[d]	400
	ISE [22]	Colorado, Pueblo, CF&I[e]	530
		Illinois, Whiteside, National[e]	400

		Pennsylvania, Bucks, U.S. Steel[e]	300
		Oregon, Portland, Oregon Steel[d]	300
	ISE [21]	Pennsylvania, Dauphin, Bethlehem	390
		Illinois, Cook, U.S. Steel	800
		Total	3120
direct reduction	McManus [27]	Oregon, Portland, Oregon Steel[d]	400000 tons per year
		South Carolina, Georgetown[f]	400000 tons per year
		Texas, Harris, Armco[f]	400000 tons per year
		West Coast, National[f]	?
		Michigan, Standard Occidental[f]	?
		Gulf Coast, Youngstown[f]	?

[a] All capacities are in terms of "tons per heat" unless otherwise specified.
[b] Not new OH installations but old OH's likely to remain until 1980.
[c] Speculation by source.
[d] Completed by January 1972.
[e] Completed by January 1973.
[f] Completed sometime after January 1972.
[g] Installations with capacities greater than 100 tons per year.

produced by two electric arc furnaces with a combined capacity of 800 tons per heat. No information is available on the expected capacity of the electric arc furnaces at Bucks county. According to the report in 33 [68], electric arc furnaces will not replace the open hearth capacity there but will supplement it. For purposes of this report, each of the two electric arc furnaces at Bucks county will be assigned a nominal capacity of 150 tons per heat.

New Technologies. Coke pelletizing is a closed system designed to eliminate gas and dust emissions that usually occur with by-product coking. Coal is heated and pelletized. In some cases the pitch removed from the gas stream is used to bind the coke particles together into a pellet. All surplus gas from the process is cleaned up for use in the steel plant. A pilot project on the feasibility of this approach is now being conducted by five

major steel and coke producers [35]. Capacities for 1980 production levels have not been published.

Direct reduction furnaces can employ lower grade iron ore and
fuels unsuitable for blast furnace iron production. This process
can produce steel directly from ore, as well as a low carbon
sponge-iron that may be used as the melting stock for steel production. The process is being considered in this latter capacity
as an alternative to scrap in the electric arc furnace. Presently
one direct reduction plant with an annual capacity of 400,000
tons is in operation in the United States [34]. Capacities are published for only four of seven direct reduction plants on stream
for 1980.

Given the data in Table 12 on particulate emissions for 1970
and the predictions in Table 13, the percentage distribution of
national particulate emissions by county was estimated for 1980.
It is presented in Table 12.

2.3 Meteorological Model: Component C

In this section, local emissions due to local coke, iron, and steel
production are related to ambient air quality concentrations
(micrograms of particulates per cubic meter of air). The concentrations are calculated by the following Gaussian plume equation
for a point source [48] and represent annual average concentrations along the centerline of the plume at ground level:

$$\chi = Q \times \frac{N \left[\exp - .5 \left(H / \sigma_z\right)^2\right] \times WT \times 313100}{\sqrt{2\pi} \times \sigma_z \times u \times (\pi/16) \times x} \tag{2}$$

where

χ = ambient air quality concentration of gas or aerosol $\leqslant$ 20
      microns in diameter (micrograms per cubic meter)

Q = uniform emissions rate (tons per year)

x = downwind distance from source (meters)

σ_z = vertical diffusion coefficient

u   = mean wind velocity affecting plume (meters per second)
WT = proportion of emissions with particle diameter $\leqslant 20$
       microns
N   = wind index
H   = effective stack height (meters)
      $= \overline{H} + \Delta H$

$$= \overline{H} + \frac{V_s \times d}{u} \left[1.5 + 2.68 \left(\frac{T_s - 283}{T_s}\right) d\right] \quad G$$

where
V_s = velocity of stack gas (meters per second)
d   = diameter of stack (meters)
u   = mean wind velocity affecting plume (meters per second)
T_s = temperature of stack gas (degrees Kelvin)
G   = scaling factor

Equation 2 is based on the following assumptions:

(1)   Plume spread has Gaussian distribution in the vertical (σ_z);
(2)   Total reflection of the plume occurs at the earth's surface
       (no fallout);
(3)   No diffusion of the plume occurs in the direction of the
       plume or perpendicular to the plume in the horizontal
       direction.

2.3.1 Acquisition of Data

In equation 2 some variables are related to local meteorological
conditions and others are related to the physical nature of the
emissions themselves. For this study meteorological data (normalized annual wind rose for sixteen wind directions and for five or
six stability classes) were obtained from the Star Program data
bank of the National Climatic Center, Asheville, North Carolina
[32]. Data were obtained for thirteen of the fifty-six counties
cited in Section 2.2.1. Each of the thirteen counties chosen
houses a single coke, iron, and steel complex. Some of the meteorological data are from weather stations in a nearby county. A list
of weather stations and counties is presented in Table 14.

Table 14. *Location of Weather Stations Used for Estimating Air Pollution in Thirteen Steel-Producing Areas*

	U.S. Weather Station			Steel Production Area	
Number	City	County	State	County	State
12906	Houston	Harris	Texas	Harris	Texas
13739	Philadelphia	Philadelphia	Penna.	Bucks	Penna.
13781	Wilmington	New Castle	Del.	New Castle	Del.
13876	Birmingham	Jefferson	Ala.	Etowah	Ala.
13874	Atlanta	Fulton	Georgia	Fulton	Georgia
13994	St. Louis	Jefferson	Mo.	Madison	Illinois
14820	Cleveland	Cuyahoga	Ohio	Lorain	Ohio
14913	Duluth	St. Louis	Minn.	St. Louis	Minn.
23174	Los Angeles	Los Angeles	Calif.	San Bernardino	Calif.
24127	Salt Lake City	Salt Lake	Utah	Utah	Utah
93721	Baltimore	Baltimore City	Md.	Baltimore City	Md.
93318	Huntington	Cabell	W. Va.	Scioto	Ohio
94830	Toledo	Lucas	Ohio	Lucas	Ohio

Because of their physical nature, particulate emissions into the atmosphere were grouped into four categories:

(1) Emissions from all coke production;
(2) Emissions from all controlled iron- and steel-making processes;
(3) Emissions from all uncontrolled but controllable iron- and steel-making processes;
(4) Emissions from all uncontrollable electric arc operations (furnace loading).

The values assigned to variables in equation 2 for characterizing each category of emissions is listed in Table 15.

2.3.2 Pollution Potential

Division of Q by X in equation 2 yields a useful ratio called "pollution potential." It measures how many tons of annual emissions are necessary to increase the annual average particulate concen-

Table 15. *Value of Variables in Gaussian Plume Equation, by Emissions Category*

Vari-able	Physical Unit	Emissions Category			
		1	2	3	4
T_s	degrees Kelvin	796	519	796	283
d	meters	8	3	3	3
H	meters	0	30.5	30.5	30.5
v_s	meters per second	1	18.6	18.6	1.0
WT		.05	1.00	1.00	.05

Source: [18].

tration in the atmosphere by a single unit. It has the units "tons of emissions per year" per "microgram per cubic meter."

Table 16 illustrates how pollution potential (Q/X) varies for the thirteen counties and for the four categories of particulate emissions. Pollution potential was computed at a distance of 2000 meters (.378 miles) from the source in the direction from the source (north through north northwest) that has the greatest value of Q/X for that locality. Note that Q/X varies among counties because of meteorological conditions and among emissions categories because of the physical characteristics of the emissions. In this study the sum of the ambient particulate concentrations for each category of emissions in a given county at a given distance from the source will be considered the local air-quality level resulting from local production activity. This approach, obviously an oversimplification, assumes that all production operations for a production complex are at a single point.

3.0 IMPLEMENTATION OF THE THREE-COMPONENT MODEL

The model described in section 2 was implemented with real data to illustrate how it can relate local air pollution levels, air pollution

Table 16. *Pollution Potential (Q/X)* by County and by Emissions Category*

County	State	Coke	Controlled Iron and Steel	Uncontrolled Iron and Steel	Uncontrollable EA
			Emissions Category		
Etowah	Ala.	49	784	118	21
San Bernardino	Calif.	14	214	304	11
New Castle	Del.	35	533	727	20
Fulton	Ga.	34	525	707	22
Baltimore City	Md.	32	563	817	15
St. Louis	Minn.	29	417	559	27
Jackson	Miss.	34	925	129	27
Lorain	Ohio	28	423	606	19
Lucas	Ohio	34	515	733	23
Scioto	Ohio	22	450	727	9
Bucks	Penna.	40	650	961	21
Harris	Texas	35	573	847	15
Utah	Utah	27	414	579	15

*Q/X: annual emissions (tons) required to increase annual average particulate concentration of particulates by one microgram at 2000 meters from point of emission.

control strategies, and national economic activity. Two sets of computations were made. The first set describes the impacts of various air pollution control strategies available to the integrated iron and steel industry on local air quality, national cross outputs, and prices of goods and services. National gross outputs of goods and services and of particulate emissions were calculated for both 1970 and 1980 final demands [56, 57] using three 102-order augmented technological matrices representing three sets of conditions for 1970.

MATRIX 1. 1970 raw steel process mix (37 percent open hearth, 49 percent basic oxygen process, and 14 percent electric arc steel) and 1970 levels of air pollution control in the steel industry.

MATRIX 2. 1970 raw steel process mix cited for matrix 1 and 100 percent utilization of air pollution control for all processes that are controllable.

MATRIX 3. Predicted 1980 raw steel process mix (13 percent open hearth, 22 percent electric, 65 percent basic oxygen process steel) and 100 percent utilization of air pollution control for all processes that are controllable.

The general equation for this set of computations was:

$$\begin{bmatrix} X_1 \\ \hline X_2 \end{bmatrix} = \begin{bmatrix} I - A_{11} & | & +A_{12} \\ \hline -A_{21} & | & I + A_{22} \end{bmatrix}^{-1} \begin{bmatrix} Y_1 \\ \hline Y_2 \end{bmatrix}$$

where X_1 = gross output of goods and services

X_2 = gross output of particulate emissions

Y_1 = final demand for goods and services

Y_2 = the unabated particulate emissions "delivered" to the public

I = identity matrix.

The matrix of technical coefficients, A, is partitioned:

$$[A] = \begin{bmatrix} A_{11} & | & A_{12} \\ \hline A_{21} & | & A_{22} \end{bmatrix}$$

where the subscript 1 refers to goods-producing sectors and subscript 2 refers to emissions and abatement activities.

For all three A matrices, national outputs of particulate emissions computed from equation 2 were translated to local emissions by county, for all counties where coke, iron, and/or steel will be produced in 1980. Local emissions in turn were translated into annual ambient particulate concentrations for those thirteen counties where meteorological data were available.

Price computations were carried out on matrices 1, 2, and 3 according to the price equation:

$$(P_1 \mid P_2) = (V_1 \mid V_2) \begin{bmatrix} I - A_{11} & +A_{12} \\ -A_{21} & I + A_{22} \end{bmatrix}^{-1}$$

where $(P_1 \mid P_2)$ is a partitioned row vector of price changes as compared with the base year. Prices of goods and services are represented by P_1 and the prices of pollution abatement are represented by P_2.

$(V_1 \mid V_2)$ is a partitioned value-added vector. Value added in goods sectors is represented by V_1 and value added in pollution abatement sectors is represented by V_2.

The above computation assumes that the cost of pollution abatement is attributed directly to the sector that uses the device. Alternatively, abatement costs for the integrated iron and steel industry could have been spread across all industries in the form of taxes or they could have been treated as paid by government subsidy. Prices of goods and services computed from matrices 2 and 3 show less than .01 percent change from prices computed from matrix 1. This change is considered insignificant.

The second set of computations relates local ambient particulate concentrations to gross output on a national level. It shows what national levels of air pollution control (and outputs of goods and services) would be required to maintain given limits on concentrations of particulates from specific steel-making processes in each particular county.

Local particulate emissions by process (q) are the quotient of ambient particulate concentration (X) at a given distance and direction from a point source divided by pollution potential, Q/X (see section 2.3.2). Allowable emissions were estimated for the thirteen counties, assuming that a maximum of 15 micrograms per cubic meter of uncontrolled particulates from electric

arc steel production and 10 micrograms per cubic meter of uncontrolled particulates from open hearth production at 2000
meters from the production site are tolerated. In order to de-
duce national particulate emissions, we sum these emissions and
assume that:

(1)   Meteorological conditions and limits on particulate
      concentrations in these thirteen counties are represen-
      tative of all steel centers in the United States;
(2)   Emissions by county are a fixed percentage of national
      emissions. From Table 11 we see that the thirteen
      counties represent 23 percent of the open hearth emis-
      sions and 14 percent of the electric arc emissions in the
      United States in 1970.

The allowable national emissions levels are now entered as negative final demands in the input-output format using the technology of matrix 2. Outputs are then calculated.

4.0 RESULTS

Using the 1970 bill of final demands [56] and matrix 1 (from
section 3), we compute the 1970 gross output of goods and
services (hereafter called standard gross output). The annual
particulate emissions assessed for the standard gross output for
the entire United States are presented in Table 17, column 1,
and by county in Table 18. Ambient particulate concentrations are estimated for thirteen of these counties, in sixteen
wind directions, at distances up to twelve miles from the
production site. In Table 19 we compare air quality concentrations in the particular wind direction associated with the
worst air pollution condition in each county. We see that
the 1970 maximum annual average particulate concentration

Table 17. *U.S. Annual Particulate Emissions:*
Four Different Assumptions*
(millions of pounds)

	Assumptions			
Emissions Sector	1970 Steel Technology, 1970 Air Pollution Control, "Standard Output"	1970 Steel Technology, Best Air Pollution Control	1980 Steel Technology, Best Air Pollution Control	1970 Steel Technology, Air Pollution Control to Yield 10 Micrograms per Cubic Meter from OH and 15 Micrograms per Cubic Meter from EA
Gross	13384	13384	13881	13384
Controlled				
Already controlled, 1970	12247	12247	13047	12247
EA controlled over 1970	0	79	114	72
OH controlled over 1970	0	461	164	459
Uncontrolled				
Coke uncontrolled	498	498	498	498
EA uncontrolled	79	0	0	7
OH uncontrolled	458	0	0	2
EA uncontrollable	19	19	27	19
Iron wet scrubbers	9	9	9	9
BOP wet scrubbers	14	14	18	14
EA fabric filters	0	0	1	0
OH electrostatic precipitators	33	73	26	73

*Final Demand constant at 1970 level.

Table 18. *Annual Particulate Emissions (Tons) for Standard Output, by County and Type of Emissions*

State	County	Uncontrollable		Controllable but Uncontrolled				Controllable and Controlled			
		By-Product Coking	Electric Arc Steel	Blast Furnace Iron	Basic Oxygen Steel	Electric Arc Steel	Open Hearth Steel	Blast Furnace Iron	Basic Oxygen Steel	Electric Arc Steel	Open Hearth Steel
Ala.	Etowah	2491.	203.	0.	0.	871.	0.	36.	142.	2.	0.
Ala.	Jefferson	20924.	0.	0.	0.	0.	10761.	288.	0.	0.	783.
Cal.	Alameda	0.	0.	0.	0.	0.	1603.	0.	0.	0.	117.
Cal.	Los Angeles	0.	185.	0.	0.	792.		0.	0.	2.	0.
Cal.	San Bernard.	6227.	0.	0.	0.	0.	4808.	76.	155.	0.	350.
Colo.	Pueblo	3986.	0.	0.	0.	0.	2747.	63.	115.	0.	200.
Del.	New Castle	0.	157.	0.	0.	673.	0.	0.	0.	2.	0.
Ga.	Fulton	0.	92.	0.	0.	396.	0.	0.	0.	1.	0.
Ill.	Cook	5978.	379.	0.	0.	1624.	12363.	337.	418.	4.	899.
Ill.	Madison	2491.	213.	0.	0.	911.	0.	49.	209.	2.	0.
Ill.	Peoria	0.	157.	0.	0.	673.	916.	0.	0.	2.	67.
Ill.	Whiteside	0.	435.	0.	0.	1861.	0.	0.	0.	5.	0.
Ind.	Howard	0.	157.	0.	0.	673.	0.	0.	0.	2.	0.
Ind.	Lake	32632.	129.	0.	0.	554.	36403.	418.	533.	1.	2647.
Ind.	Porter	2242.	0.	0.	0.	0.	0.	49.	283.	0.	0.
Ky.	Campbell	0.	83.	0.	0.	356.	1832.	0.	0.	1.	133.
Ky.	Daviess	0.	65.	0.	0.	277.	0.	0.	0.	1.	0.

State	County	Uncontrollable		Controllable but Uncontrolled				Controllable and Controlled			
		By-Product Coking	Electric Arc Steel	Blast Furnace Iron	Basic Oxygen Steel	Electric Arc Steel	Open Hearth Steel	Blast Furnace Iron	Basic Oxygen Steel	Electric Arc Steel	Open Hearth Steel
Md.	Baltimore City	14946.	92.	0.	0.	396.	13966.	238.	189.	1.	1016.
Mich.	Wayne	11957.	268.	0.	0.	1148.	0.	459.	972.	3.	0.
Minn.	St. Louis	2242.	0.	0.	0.	0.	3663.	27.	0.	0.	266.
Mo.	Jackson	0.	296.	0.	0.	1267.	0.	0.	0.	3.	0.
N.J.	Burlington	0.	65.	0.	0.	277.	0.	0.	0.	1.	0.
N.Y.	Erie	13451.	102.	0.	0.	436.	8242.	238.	418.	1.	599.
N.Y.	Niagara	0.	0.	0.	0.	0.	0.	9.	0.	0.	0.
N.Y.	Rensselaer	0.	0.	0.	0.	0.	0.	13.	0.	0.	0.
Ohio	Butler	3487.	305.	0.	0.	1307.	4808.	63.	189.	3.	350.
Ohio	Cuyahoga	6975.	157.	0.	0.	673.	4121.	225.	432.	2.	300.
Ohio	Jefferson	6227.	0.	0.	0.	0.	6182.	94.	283.	0.	450.
Ohio	Lorain	4733.	0.	0.	0.	0.	5953.	108.	0.	0.	433.
Ohio	Lucas	2989.	0.	0.	0.	0.	4121.	31.	0.	0.	300.
Ohio	Mahoning	9466	0.	0.	0.	0.	13966.	85.	0.	0.	1016.
Ohio	Richland	0.	111.	0.	0.	475.	2518.	0.	0.	1.	183.
Ohio	Scioto	1245.	0.	0.	0.	0.	0.	49.	0.	0.	0.

State	County										
Ohio	Stark	498.	990.	0.	0.	4237.	0.	27.	0.	11.	0.
Ohio	Trumbull	1495.	444.	0.	0.	1901.	0.	216.	142.	5.	0.
Pa.	Allegheny	35372.	712.	0.	0.	3049.	35945.	490.	358.	8.	2614.
Pa.	Armstrong	5231.	0.	0.	0.	0.	0.	0.	0.	0.	0.
Pa.	Beaver	7722.	527.	0.	0.	2257.	0.	171.	378.	6.	0.
Pa.	Bucks	3487.	0.	0.	0.	0.	9387.	99.	0.	0.	683.
Pa.	Cambria	6227.	777.	0.	0.	3326.	0.	108.	688.	8.	0.
Pa.	Chester	0.	185.	0.	0.	792.	4350.	0.	0.	2.	316.
Pa.	Dauphin	0.	250.	0.	0.	1069.	2289.	0.	0.	3.	166.
Pa.	Erie	0.	102.	0.	0.	436.	0.	0.	0.	1.	0.
Pa.	Lawrence	0.	0.	0.	0.	0.	458.	0.	0.	0.	33.
Pa.	Mercer	0.	120.	0.	0.	515.	13737.	306.	142.	1.	999.
Pa.	Mifflin	0.	92.	0.	0.	396.	0.	0.	0.	1.	0.
Pa.	Montgomery	2989.	0.	0.	0.	0.	229.	22.	135.	0.	17.
Pa.	Northampton	7473.	129.	0.	0.	554.	0.	112.	236.	1.	0.
Pa.	Westmorel.	1744.	0.	0.	0.	0.	0.	54.	0.	0.	0.
S.C.	Georgetown	0.	65.	0.	0.	277.	0.	0.	0.	1.	0.
Tenn.	Hamilton	747.	0.	0.	0.	0.	0.	0.	0.	0.	0.
Tex.	Harris	996.	444.	0.	0.	1901.	1832.	0.	0.	5.	133.
Tex.	Morris	1495.	0.	0.	0.	0.	3205.	27.	0.	0.	233.
Utah	Utah	4982.	0.	0.	0.	0.	8929.	76.	0.	0.	649.

State	County	Uncontrollable		Controllable but Uncontrolled				Controllable and Controlled			
		By-Product Coking	Electric Arc Steel	Blast Furnace Iron	Basic Oxygen Steel	Electric Arc Steel	Open Hearth Steel	Blast Furnace Iron	Basic Oxygen Steel	Electric Arc Steel	Open Hearth Steel
W. Va.	Marion	996.	0.	0.	0.	0.	0.	0.	0.	0.	0.
W. Va.	Hancock	6477.	0.	0.	0.	0.	5724.	103.	317.	0.	416.

Table 19. *Estimated Ambient Particulate Concentrations Due to Coke, Iron, and/or Steel Production in 1970**

County	State	Miles from Source								
		.09	.16	.31	.47	.62	1.24	3.11	6.21	12.43
		Concentration (micrograms per cubic meter)								
Etowa	Ala.	17.9	24.3	24.8	22.2	19.9	12.6	4.5	1.8	1.0
San Bernard.	Cal.	182.6	281.3	346.6	334.8	314.0	207.6	70.3	26.4	11.6
New Castle	Del.	2.9	4.3	9.6	12.9	13.7	9.7	3.3	1.2	0.7
Fulton	Ga.	2.0	2.8	5.9	7.7	8.1	5.8	1.9	0.7	0.4
Baltimore	Md.	164.2	253.1	348.1	362.1	342.9	224.7	86.9	40.9	28.7
St. Louis	Minn.	27.2	43.6	67.1	94.5	104.8	76.4	23.9	8.2	3.7
Jackson	Mo.	4.5	5.9	11.7	13.6	13.6	10.4	3.8	2.2	1.8
Lorain	Ohio	43.4	76.5	113.6	138.6	152.9	119.6	41.2	15.4	7.4
Lucas	Ohio	27.2	46.6	67.9	84.2	90.5	67.0	22.9	8.5	3.9
Scioto	Ohio	10.0	13.6	12.8	9.8	7.7	3.9	1.3	0.5	0.4
Bucks	Pa.	72.9	98.1	112.5	102.7	91.1	56.6	21.4	9.5	5.3
Harris	Tex.	22.8	33.8	63.3	68.9	67.7	49.0	17.8	6.8	2.7
Utah	Utah	68.1	118.8	196.3	241.0	252.2	181.1	60.9	23.3	13.6

*By county and distance from production site, resulting from 1970 "Standard Output" discussed in text.

Table 20. *Changes in Gross Output from 1970 Standard to 1980**

A. Change in Raw Steel Production Mix

Sector	Millions of 1958 Dollars	Sector	Millions of 1958 Dollars
Iron and ferroalloy ores mining (5)	-9.1	Coke (39[S])	21.9 MT
Coal mining (7)	1.8	Electric arc steel (40[S])	8.9 MT
Crude petroleum and natural gas (8)	-6.2	Open hearth steel (41[S])	-30.9 MT
Chemicals and fertilizer mining (10)	2.1	(43[S])	8.9 MT
Food and kindred products (14)	1.3	Primary nonferrous metals manufacturing (38)	4.9
Broad and narrow fabrics yarn and thread (16)	2.3	Screw machine prods. (41)	1.0
Paper and allied prod. ex. containers & boxes (24)	2.3	Other fabricated metal products (42)	1.2
Paper cont. & boxes (25)	1.1	Electric gas, water and sanitary serv. (68)	41.3
Printing and publ. (26)	2.0	Wholesale and retail trade (69)	3.3
Chemicals and selected chemical products (27)	50.4	Finance and insurance (70)	1.3
Plastics and synthetic materials (28)	2.3	Business services (73)	2.2
Petroleum refining and related indust. (31)	-17.9	State and local gov't. enterprises (79)	5.1
Rubber and misc. plastics products (32)	1.2	Business travel entertainment and gifts (81)	1.3
		Scrap, used and secondhand goods (83)	18.6

B. Change in Raw Steel Production Mix and 20 Percent Increase in Imported Semi-Finished Steel

Sector		Sector	
Livestock (1)	-1.4	Crude petroleum and natural gas (8)	-20.5
Other agric. prods. (2)	-2.0	Stone and clay mining and quarrying (9)	-1.4
Iron and ferroalloy ore mining	-145.0	Chemicals and fertilizer mining (10)	-1.3
Nonferrous metal ores mining (6)	-5.8	Maint. & repair constr. (12)	-8.0
Coal mining (7)	-57.8		

Food & kindred prods. (14)	-2.8	Screw machine prods. (41)	-5.6
Broad and narrow fabrics, yarn and thread (16)	-9.2	Other fabr. metal prod. (42)	-8.7
Lumber and wood prods. ex. containers (20)	-3.2	Constr., mining, oil field mach. and equip. (45)	-8.6
Paper and allied prods. ex. containers & boxes (24)	-8.1	Metalworking mach. and equipment (47)	-2.3
Paperboard containers and boxes (25)	-3.6	Spec. indust. mach. & equip. (48)	-1.6
Printing and publishing (26)	-8.7	General ind. mach. & equip. (49)	-4.1
Chemicals & sel. chem. prod. (27)	-26.6	Mach. shop prods. (50)	-4.0
Plastics and synthetic materials (28)	-4.8	Elec. transmission and distr. equip. (53)	-3.8
Drugs, cleaning and toilet prep. (29)	-1.5	Radio, TV, commun. equip.	-5.4
Paints and misc. plastics prods. (30)	-1.0	Elec. comp. and access. (57)	-2.9
Petroleum refining and related industry (31)	-42.4	Motor vehicles & equip. (59)	-15.0
Rubber and misc. plastics products (32)	-6.0	Aircraft & parts (60)	-1.5
Stone and clay prods. (36)	-9.1	Other transp. equip. (61)	-2.4
Coke (37^S)	-5.6 MT	Prof. sci. & contr. instr. (62)	-1.3
Blast furnace iron (38^S)	-8.9 MT	Misc. manuf. (64)	-1.8
Basic oxygen steel (39^S)	14.9 MT	Transp. & warehousing (65)	-39.5
Electric arc steel (40^S)	6.8 MT	Comm. ex. radio, TV (66)	-3.8
Open hearth steel (41^S)	-35.8 MT	Wholesale & ret. trade (69)	-15.9
Home scrap (42^S)	-3.6 MT	Finance and insurance (70)	-8.5
(43^S)	-209.3	Real est. and rental (71)	-21.8
(44^S)	-7.0	Hotels, lodg., pers. repair ex. auto (72)	-1.3
Prim. nonferrous metals manuf. (38)	-22.4	Business serv. (73)	-12.4
Metal containers (39)	-1.1	Auto & repair serv. (75)	-2.0
Heating, plumbing and fabricated (40)	-2.9	Fed. & gov't. enterp. (78)	-1.2
		State & local gov't. (79)	-1.5
		Gross imports of G & S (80)	73.9
		Bus. travel, entertain, gifts (81)	-4.4
		Scrap, used, and 2nd hand goods (83)	-75.0

*Final demand constant at 1970 level. Changes recorded are greater than one million 1958 dollars (or one million tons, where indicated).

ranged from eight to 362 micrograms per cubic meter at .3 to
.4 miles from the production site. At distances of six miles, in the
same wind direction, concentrations ranged from one to 41 micrograms per cubic meter.

Keeping final demand and production technology at the 1970
levels, we now introduce 100 percent application of best feasible
air pollution control (matrix 2). This induces two sectors to increase their annual output by more than $1 million (maintenance
and repair construction, $1.4 million; utilities, $2.8 million).
(See Table 20.) It also means a decrease in annual national emissions of 540 million pounds. An analysis is given in Table 17.
Because local process mix differs, however, the percentage of
reduction in emissions is not uniform in all counties. This is
shown in Table 21. In Figures 3–5 we illustrate the effects of
controls on three counties: San Bernardino, California; Bucks,
Pennsylvania; and Etowah, Alabama.

Only areas producing electric arc and open hearth steel show
decreases in particulate levels as controls are imposed. Furthermore, the percentage of reductions in particulate levels for a
single locality are not uniform at all distances from the production site. Each of the four categories of emissions cited in Table
17 has unique characteristics affecting the contributions of emissions to air pollution at specific distances from the source. When
additional air pollution control devices are fitted onto electric
arc and open hearth furnaces, air pollution levels greatly decrease
from the 1970 levels at distances far from the source. The decrase
is smaller, however, as we approach the source. And in some
areas, air pollution levels are even greater than the 1970 levels
when measured very close to the source. This increase is due to
the physical characteristics of the emissions from the air-cleaning
device itself.

Introducing the 1980 process mix along with 100 percent
application of best feasible control technology brings further
departures from the standard gross output and a reduction of
1078 million pounds from the emissions calculated for the
standard gross output. With the 1980 process mix, improvements

Table 21. *Percent Improvement Over Base Standard Air Quality With Introduction of Best Available Air Pollution Controls Assuming 1970 Final Demand (Base standard Air Quality Levels are given in Table 18)*

A. 1970 Steel Process Mix

County	State	Miles from source								
		Percent Improvement								
		.09	.16	.31	.47	.62	1.24	3.11	6.21	12.43
Etowah	Ala.	0.6	5.3	21.9	34.5	43.6	58.7	60.2	45.8	33.1
San Bernard.	Cal.	-1.1	11.6	36.9	47.4	54.6	66.7	70.2	66.6	65.5
New Castle	Del.	3.3	36.1	80.9	91.1	94.2	95.5	91.1	89.2	92.7
Fulton	Ga.	3.5	38.1	81.6	91.1	94.3	95.9	92.0	90.5	93.8
Baltimore	Md.	-1.0	10.1	39.3	52.4	59.3	68.6	64.4	51.7	54.7
St. Louis	Minn.	-0.6	3.1	31.4	53.9	64.2	75.8	79.3	75.1	73.1
Jackson	Mo.	3.2	38.8	82.2	90.7	93.7	94.3	87.0	89.7	95.1
Lorain	Ohio	-2.3	3.1	28.8	45.9	57.5	72.0	75.1	70.2	67.0
Lucas	Ohio	-0.8	6.8	33.9	51.7	61.4	73.5	76.7	72.6	68.6
Scioto	Ohio	0.0	0.0	0.0	0.0	0.0	0.0	0.0	0.0	0.0
Bucks	Pa.	0.7	9.3	33.0	44.1	50.6	61.0	58.3	49.8	50.1
Harris	Tex.	0.1	21.1	62.6	72.6	77.5	84.2	83.1	78.7	70.6
Utah	Utah	-0.6	13.0	42.3	56.7	64.4	74.9	76.8	68.3	66.6

B. 1980 Steel Process Mix

County	State	Percent Improvement								
Miles from source		.09	.16	.31	.47	.62	1.24	3.11	6.21	12.43
Etowah	Ala.	-0.6	4.6	20.6	33.1	42.2	57.2	57.9	43.3	31.4
San Bernard.	Cal.	5.9	26.3	55.1	67.3	74.3	83.6	84.3	81.7	82.6
New Castle	Del.	-23.2	18.5	75.5	88.5	92.5	94.2	88.5	86.1	90.6
Fulton	Ga.	-11.5	28.3	78.6	89.6	93.3	95.1	90.7	88.9	92.7
Baltimore	Md.	2.8	14.7	43.7	56.4	63.0	71.6	67.1	55.3	58.7
St. Louis	Minn.	3.6	13.2	60.0	81.7	88.6	94.3	94.0	92.0	92.8
Jackson	Mo.	-17.9	25.3	78.1	88.6	92.2	93.0	84.2	87.3	94.0
Lorain	Ohio	10.0	16.7	51.5	70.4	79.6	88.6	88.7	85.4	85.2
Lucas	Ohio	5.5	20.4	59.6	78.2	85.8	92.7	92.4	89.6	89.3
Scioto	Ohio	1.9	1.9	1.7	1.6	1.5	1.4	1.3	1.3	1.6
Bucks	Pa.	39.7	21.4	17.3	18.1	19.8	29.5	39.7	32.2	19.6
Harris	Tex.	-36.0	12.9	69.7	82.3	87.7	92.5	82.1	76.8	73.9
Utah	Utah	-2.9	4.6	28.4	42.7	51.6	64.8	68.4	58.2	54.3

C. 1980 Steel Process Mix, 20 Percent Increase in Steel Imports

Etowah	Ala.	-1.6	3.7	20.1	32.8	42.0	57.2	57.8	43.1	31.0
San Bernard.	Cal.	4.6	25.3	54.6	67.0	74.1	83.5	84.2	81.5	82.5
New Castle	Del.	-22.3	19.3	76.0	88.8	92.8	94.4	88.7	86.4	90.9
Fulton	Ga.	-10.6	29.1	79.0	89.8	93.5	95.3	90.8	84.9	90.0
Baltimore	Md.	1.5	13.8	43.3	56.2	62.9	71.6	66.9	55.0	58.6
St. Louis	Minn.	2.2	11.9	59.5	81.5	88.5	94.2	93.9	91.9	92.7
Jackson	Mo.	-17.0	26.1	78.5	88.8	92.5	93.2	84.4	87.6	94.2
Lorain	Ohio	8.8	15.5	51.0	70.1	79.4	88.5	88.6	76.6	76.3
Lucas	Ohio	4.1	19.3	59.1	78.0	85.6	92.6	92.3	89.5	89.1
Scioto	Ohio	0.6	0.6	0.6	0.7	0.7	0.7	0.7	0.7	0.7
Bucks	Pa.	39.2	21.2	17.3	18.2	20.0	29.7	39.8	32.3	19.7
Harris	Tex.	-35.7	13.0	69.8	82.5	87.8	92.6	82.2	77.0	74.0
Utah	Utah	-4.1	3.8	28.1	42.6	51.6	64.8	68.4	58.2	54.4

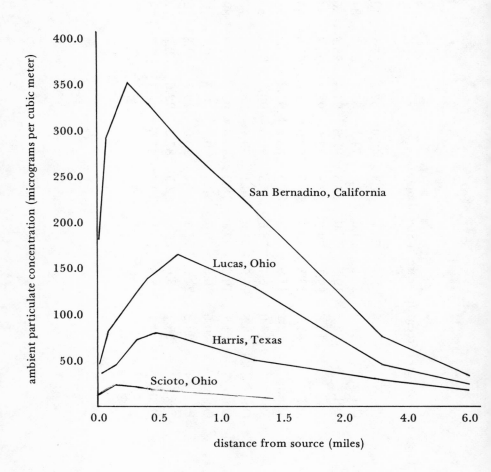

Figure 3. *Contributions from Coke, Iron and Steel Production to Air Pollution, 1970: San Bernardino County, California*

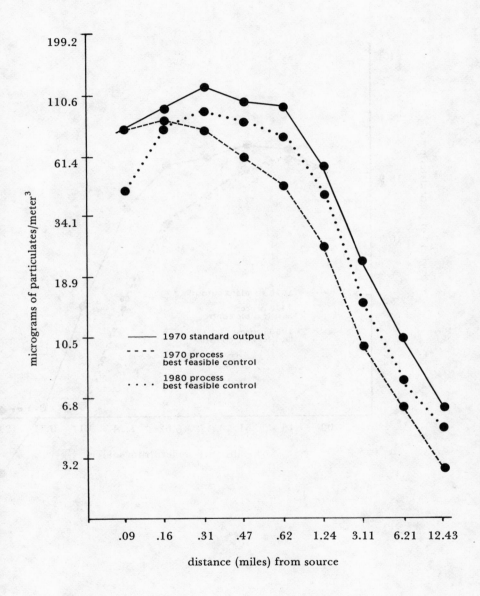

Figure 4. *Contributions from Coke, Iron and Steel Production to Air Pollution, 1970: Bucks County, Pennsylvania*

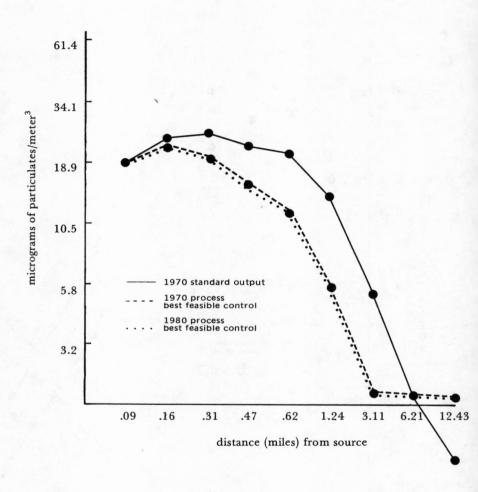

Figure 5. *Contributions from Coke, Iron and Steel Production to Air Pollution, 1970: Etowa County, Alabama*

in air quality are greater where open hearth furnaces will be eliminated than where they will remain. This is because open hearth production will be diminished by 1980; but where open hearth furnaces remain, production will be relatively greater than in 1970.

In the calculations described above, final demand is kept constant at the 1970 level. The effect of switching to 1980 final demand is also estimated. Depending on which of the two versions of the projected 1980 final demand [57] we chose, the emissions levels will be 59–67 percent greater than with the 1970 final demand.

In our last set of computations we set local air quality standards and then determined what these standards would mean for national particulate emissions. First, we calculated the maximum permissible emissions in thirteen counties if the allowable concentrations were 10 micrograms per cubic meter for open hearth uncontrolled emissions and 15 micrograms per cubic meter for electric arc uncontrolled emissions. We now estimate national emissions, assuming that the emissions from the thirteen counties are the fixed percentage of national emissions given in Table 12. Our air quality standards imply that the nation would accept no more than a total of 1022 tons of uncontrolled open hearth and 3694 tons of uncontrolled electric arc particulate emissions. Pollution abatement required to keep within these tolerances would induce a small change in gross output as compared to the standard gross output. Particulate emissions levels would be 521 million pounds less than those calculated for the standard gross output.

5.0 SUMMARY AND CONCLUSIONS

In this study a flexible ecologic-economic model relating local pollution to national economic parameters is introduced and applied to the integrated iron and steel industry in the United

States. The model shows that significant decreases in particulate concentrations in the ambient air occur when air pollution controls are imposed on currently uncontrolled facilities. Imposition of controls does induce changes in prices and in gross output of goods and services, but the changes are small.

The flexibility of this model is important for two reasons. First, it allows us to set local air quality levels by process and then derive the effect on the economy or, conversely, to fix the national final demand on output levels and then determine local air quality. Second, the model is easy to modify and update with respect to the number of industries, localities, and pollutants being assessed. To make this model more useful as a planning tool, we suggest that future work be focused on the following areas.

(1) *Refinement of data and methodology.* Three assumptions made in this study should be reviewed for their validity. First, it is assumed that the generalized engineering parameters for the entire iron and steel industry are applicable to each facility. It is also assumed that the simple Gaussian plume model for estimating air quality levels is applicable for all localities, regardless of topography. Furthermore, only local emissions directly related to coke, iron, and steel production capacity in each county are considered. Emissions from other industries are not taken into account.

(2) *Capital goods.* Capital goods were treated as exogenous in this model. A dynamic model would take account of requirements for specific capital goods for capacity expansion and for increased abatement.

(3) *Trade-offs between environmental contaminants.* In order to assess the entire impact of air pollution controls, it is necessary to measure their effect on other sectors of the environment. For example, the effect of air cleaning devices on noise and solid waste levels should be evaluated, as well as the effect of wet scrubbers on water pollution. These interrelationships should be quantified in the A_{22} quadrant of the matrix.

(4) *Pollution control in the entire economy.* We have shown

that the effect on prices and gross output of the national economy is small, when controls are imposed on the steel industry alone. However, larger effects will appear as pollution controls are added in other industries as well.

(5) *Local air pollution.* In this study we relate local air pollution to local emissions by setting the allowable concentration for pollutants from each process. A model should be developed for setting the combined concentration for the entire region rather than for each process. Given such constraints as air pollution control costs by process, economic and population growth, zoning, etc., linear programming techniques could be used to determine the level of pollution abatement for each process in the region.

(6) *Assessment of coefficients.* After a check of the literature, we see that the air quality levels predicted by this study are within reasonable limits of measured levels. More thorough statistical analysis should be pursued, however, to assess the reliability of the predictions.

REFERENCES

1. Allegheny County Air Pollution Bureau staff, Pittsburgh, Pa. (January 1972), personal communication.
2. American Iron and Steel Institute, *Annual Statistical Report* (Washington, D.C.: American Iron and Steel Institute, 1970).
3. —— *Directory* (Washington, D.C.: American Iron and Steel Institute, 1970).
4. —— *Steel-making Flowcharts* (Washington, D.C.: American Iron and Steel Institute, 1970).
5. Anderson, W., Environmental Quality Control, Bethlehem Steel Corporation (March 1972), personal communication.
6. Barnes, T., *Evaluation of Process Alternatives to Improve Control of Air Pollution from Coke* (Columbus, Ohio: Battelle Memorial Institute, January 1970).
7. Blaskowski, H., and Sefcik, A., "Economics of Gas Cooling and Gas Cleaning Systems Associated with the BOP Process," *Combustion* (November 1967), pp. 31–35.

8. Bramer, H., "Pollution Control in the Steel Industry," *Environmental Science and Technology*, 5 (1971), 1004–08.

9. —— Datagraphics, Inc., Pittsburgh, Pa. (January 1972), personal communication.

10. Brandt, A., "Current Status and Future Prospects—Steel Industry Air Pollution Control," *Proceedings, Third National Conference on Air Pollution Control* (Washington, D.C.: Air Pollution Control Association, December 12-14, 1966), pp. 236–241.

11. Calenza, G. J., "Air Pollution Problems Faced by the Iron and Steel Industry," *Plant Engineering* (April 30, 1970), pp. 60–63.

12. Carter, A. P., "A Mathematical Model for Environmental Quality," paper presented at Primer Seminario Centroamericano sobre el Medio Ambiente, Fisico y el Desarrollo (Antigua, Guatemala, July 25–30, 1971).

13. "Continuous Steel?" *Scientific American*, 219 (1968), 46.

14. Demaree, A., "Steel. Recasting an Industry," *Fortune* (March 1971), pp. 74–77+.

15. deMenil, G., "Projections of Input Structure for the Iron and Steel Manufacturing Industry," unpublished paper (Cambridge, Massachusetts: Harvard Economic Research Project, July 1964).

16. Dunlap, R., et al., *A Study of Air Pollution Control for Allegheny County, Pa.* (Pittsburgh, Pa.: Carnegie-Mellon Institute, November 1970).

17. Ernst and Ernst, *A Rapid Cost Estimating Method for Air Pollution Control Equipment* (Washington, D.C.: Ernst and Ernst, September 1968).

18. First, M., Professor of Environmental Engineering, Harvard University (May 1972), interview.

19. Fogel, M. E., et al., *Final Report—Comprehensive Economic Cost Study of Air Pollution Control Costs for Selected Industries and Selected Regions* (North Carolina: Research Triangle Institute, February 1970), Appendix A.

20. Franklin Research Laboratory, *Industrial Guides for Control of Integrated Iron and Steel Mill Emissions and Secondary Ferrous Pyrometallurgical Emissions* (Philadelphia, Pa.: Franklin Research Institute, 1970).

21. Fritz, W. E., Division of Compliance, State of Maryland Bureau of Air Quality Control (January 1972), personal communication.

22. Glover, W., "Changing Considerations in the Design of Large Blast Furnaces," *Iron and Steel Engineer* (June 1971), pp. 56–58.

23. Hogan, W. T., "A Challenge to World Steel: Raw Materials for One Million Tons of Output," Steel Service Center Institute, 5 (May 1970).

24. —— "As Steel Closes the Books," Steel Service Center Institute, 5 (December 1969).

25. —— *The Economic History of the Iron and Steel Industry* (5 vols., Lexington, Massachusetts: Heath-Lexington Books, 1971).

26. *Iron and Steel Engineer* (January 1970).

27. *Iron and Steel Engineer* (January 1971).

28. Jablin, R., "Environmental Control at Alan Wood: Technical Problems, Regulations and New Procedures," *Iron and Steel Engineer* (July 1971), pp. 58–65.

29. Klauss, P., Suffert, P., Skelly, J., *Final Economic Report on Cost Analysis for a Systems Analysis Study of the Integrated Iron and Steel Industry* (Pittsburgh, Pa.: Swindell-Dressler Co., Contract #PH-22-68-65, May 15, 1969), Appendix C.

30. Leontief, W., "Environmental Repercussions and the Economic Structure: An Input-Output Approach," *Review of Economics and Statistics,* 52 (1970), 262–271.

31. —— and Ford, D., "Air Pollution and the Economic Structure: Empirical Results of Input-Output Computations," *Input-Output Techniques,* ed. A. P. Carter and A. Brody (Amsterdam: North-Holland Publishing Company, 1972).

32. McGraw, M. and Duprey, R., *Air Pollution Emissions Factors* (Washington, D.C.: Government Printing Office, 1971).

33. McManus, G., "Gun Goes Off for Direct Reduction," *Iron Age* (August 27, 1970), pp. 69–76.

34. "Mini-Plants in the U.S.: Robin Steel's Two Joins Forty-Two and the Number Still Grows," 33 (March 1971), 56–59.

35. "New Coking Process May End Pollution," *Bethlehem Review* (September 1971), p. 12.

36. Oglesby, S., *A Manual of Electrostatic Precipitator Technology,* Part II, Application Areas (Birmingham, Alabama: Southern Research Institute, August 25, 1970).

37. Schueneman, J., *Air Pollution Aspects of the Iron and Steel Industry* (Washington, D.C.: Public Health Service Publication No. 999-AP-1).

38. Sebesta, W., "Ferrous Metalurgical Process," *Air Pollution,* ed. A. Stern, 2nd ed., 3 (New York, N.Y.: Academic Press, 1968).

39. Shannon, L. J., et al., *Particulate Pollutant Study* (Kansas City, Missouri: Midwest Research Institute, August 1971), Vol. 2.

40. Skinner, W. and Rogers, D., *Manufacturing Policy in the Steel Industry* (Homewood, Illinois: Richard Irwin, Inc., 1970).

41. "Steel," in *McGraw-Hill Encyclopedia of Science and Technology,* 13 (New York, N.Y.: McGraw-Hill Publishing Company, 1971).

42. Stone, J., "Oxygen Steel-Making," *Scientific American,* 218 (April 1968), 24–31.

43. Study of Critical Environmental Problems (SCEP), *Man's Impact on*

the *Global Environment* (Cambridge, Massachusetts: MIT Press, 1970).

44. Tietig, R., Jr., and Kuhl, R., "Predicting Changes in Steel-Making Processes," *Iron and Steel Engineer* (June 1970), 77–85.

45. Tihansky, D., "A Cost Analysis of Waste Management in the Steel Industry," *Journal of the Air Pollution Control Association*, 22 (May 1972), 335–41.

46. —— *Patterns of Energy Demand in Steel Making* (Santa Monica, California: Rand Corporation, May 1971, WN-7437-NSF).

47. Tsao, C. and Day, R., "A Process Analysis Model of the U.S. Steel Industry," *Management Science*, 17 (June 1971), B588–B608.

48. Turner, D., *Workbook of Atmospheric Dispersion Estimates* (Washington, D.C.: Government Printing Office, 1969).

49. United Nations, *Problems of Air and Water Pollution Arising in the Iron and Steel Industry* (New York, N.Y., 1970, SE/ECE/STEEL 32).

50. U.S. Congress, *The Economics of Clean Air: Annual Report of the Administrator of the Environmental Protection Agency to the Congress of the United States* (Washington, D.C.: Government Printing Office, 1972.

51. U.S. Department of Commerce, Bureau of the Census, *1958 Census of Manufactures, Vol. 3, Industry Statistics, Part 1, Major Groups 20-28* (Washington, D.C.: Government Printing Office, 1961).

52. —— *1958 Census of Manufactures, Vol. 2, Industry Statistics, Part 2, Major Groups 29-39* (Washington, D.C.: Government Printing Office, 1961).

53. —— *Statistical Abstract of the United States,* 79th ed. (Washington, D.C.: Government Printing Office, 1958).

54. —— Office of Business Economics, *Input-Output Structure of the U.S. Economy: 1963,* 3 vols. (Washington, D.C.: Government Printing Office, 1969).

55. —— Bureau of Labor Statistics, *1970 Input-Output Coefficients,* BLS Report 326 (Washington, D.C.: Government Printing Office, 1967).

56. —— *Patterns of U.S. Economic Growth,* BLS Bulletin 1672 (Washington, D.C.: Government Printing Office, 1970).

57. —— *Projections of the Post Vietnam Economy to 1975, Appendix A,* BLS Bulletin 1733 (Washington, D.C.: Government Printing Office, 1972).

58. —— *Wholesale Price Index* (Washington, D.C.: Government Printing Office, 1972).

59. U.S. Department of Commerce, National Oceanic and Atmospheric Administration, Environmental Data Service, *Star Output Normalized Data* (Asheville, North Carolina: National Climatic Center), computer output.

60. U.S. Department of Health, Education and Welfare, Public Health Service, Environmental Health Service, National Air Pollution Control Association, *Characteristics of Particulate Patterns 1957–1966* (Washington, D.C.: Government Printing Office, March 1970).

61. U.S. Department of Interior, Bureau of Mines, *Minerals Yearbook*, Vols. 1–2 (Washington, D.C.: Government Printing Office, 1958).

62. —— Division of Atmospheric Surveillance, *Air Quality Data for 1967, Revised for 1971* (Research Triangle Park, N.C.: U.S. Environmental Protection Agency, 1971).

63. Vandegrift, A. E., et al., "Particulate Pollution in the United States," *Journal of the Air Pollution Control Association,* 21 (1971), 321–328.

64. Varga, J., *A Systems Analysis Study of the Integrated Iron and Steel Industry* (Columbus, Ohio: Battelle Memorial Institute, May 1969).

65. Walker, A., and Brown, R., *Statistics on Utilization, Performance and Economics of Electrostatic Precipitators for Control of Particulate Air Pollution* (Bound Brook, N.J.: Research Cottrell, 1970).

66. "Whither Goest Thou, U.S. Steel Industry? Steel Shipment Growth," 33 (April 1971), 30–35.

67. "Whither Goest Thou, U.S. Steel Industry? Crude Steel Production," 33 (May 1971), 50–55.

68. "Whither Goest Thou, U.S. Steel Industry? Production Process," 33 (June 1971), 38–43.

69. "Whither Goest Thou, U.S. Steel Industry? Import-Export," 33 (July 1971), 30–33.

7

Coefficients for an
Input-Output Pollution
Model from Engineering Data

TERRY JENKINS

1.0 INTRODUCTION

This study shows how engineering estimates of industrial
particulate emission rates and control costs may be used to
arrive at coefficients for an input-output model that measures
the impact of environmental problems. Particulates are only
one of several types of pollutants treated in input-output
models. For each pollutant to be dealt with, a row and a column
are added to the familiar input-output system. A row of emissions coefficients shows how much of a particular pollutant is
emitted per unit of output by each sector. Input coefficients
for the abatement sector show the purchases of current and
capital account inputs required to abate a unit of pollution.
Current account replacement and capital coefficients must be
estimated. Engineering information is used because of the scarcity of statistical data on emissions and abatement costs. Most
industrial users of abatement equipment have no need for the
kind of information that would yield input-output coefficients
and therefore very little detailed information on abatement
costs is available.

The organization of this chapter mirrors the actual procedure followed in converting engineering estimates to input-output coefficients. To provide background, section 2 presents a general discussion of pollution economics. Section 3 demonstrates that emission and abatement coefficients are functions of engineering parameters. Sections 4–6 describe the data and methods used to evaluate emission rates, abatement costs, and abatement technologies. Final estimates of the input-output coefficients appear in section 7.

2.0 AIR POLLUTION ECONOMICS

2.1 Particulate Air Pollutants

As the name suggests, particulates are solid particles or liquid droplets in airborne suspension. They range in diameter from less than one to more than 100 microns (one micron equals 10^{-6} meters). The category includes a vast array of chemically distinct bits of matter, ranging from cotton lint to iron oxide dust. Recently the National Air Pollution Control Administration (NAPCA) listed the sources of particulate emissions in the United States in 1966 [10].

Industrial (52%)
Power generation (26%)
Incineration (9%)
Space heating (9%)
Mobile (4%)

This study covers only industrial and power-generating sources and part of space heating. Together they represent approximately 80 percent of the nation's particulate problem.

Although particulates are only one of about ten major air pollutants, the physical and chemical characteristics of these dust specks make them the most easily and therefore the most widely controlled of any important effluent in the United States today. Highly refined particulate collection technologies have been developed and extensive technical literature is available on the subject. But major studies of particulate generation and control have not had a national focus until recently, with the passage of the Clean Air Acts of 1967 and 1970. Engineering research in this field ordinarily provides an in-depth view of one plant rather than industry-wide estimates of important scientific and economic parameters. Estimates of average or representative costs of abatement are almost nonexistent.

2.2 Industrial Sources of Particulate Pollution

Table 1 subdivides industrial and power-generation processes into twenty-nine principal sources of particulate pollution. All twenty-nine involve the processing of a raw material and/or fuel combustion. Particulates released by the service sector come mainly from industrial space-heating operations. This is included under the heading "other fuel combustion" in Table 1. An asterisk identifies processes for which "good" emission-rate and abatement-cost data were available. Good data are estimates we judged to be informed rather than speculative. We based our evaluations on (1) whether the estimates referred to a specific polluting process rather than a large class of processes, and (2) the reliability of the data source. By these standards, we found good information for eleven polluting processes. These eleven processes are responsible for almost two thirds of all industrial and power-plant dust, or about one half of all particulate emissions in the United States. Sections 4 and 6 rate the quality of data for the other fifteen processes listed in Table 1. We list calculations based upon high- and low-quality data separately, to permit refinement of the coefficients as better information becomes available.

Table 1. *Percentage Distribution of Industrial Raw Particulate Emissions in the United States, by Source (1968)*

Process		Raw Emissions percent of total
Fuel combustion		42.93
coal-burning electric utilities	34.01*	
other fuel combustion	8.92	
Steelmaking		13.07
blast furnaces	7.68*	
basic oxygen furnaces	1.46*	
open hearth furnaces	.74*	
sintering	.67*	
electric arc furnaces	.11*	
other steel processes	2.41	
Cement		10.31
wet process	4.82*	
dry process	3.43*	
other cement processes	2.06	
Crushed stone, sand, gravel		7.72
Asphalt		6.69
paving materials	6.66*	
other asphalt processes	.03	
Forest products		3.95
pulp-mill recovery boilers	2.41*	
other pulp and paper processes	1.26	
other forest products	.28	
Nonferrous metals		3.43
Agricultural processes		3.37
Fertilizer and phosphate rock		3.15
Lime		2.55
rotary kilns	2.05*	
other lime processes	.50	
Clay products		1.61
Ferroalloys		.48
Iron foundries		.26
Acids		.18
Coal cleaning		.12
Carbon black		.12
Petroleum		.06

*"Good" data source (see text).
Source: computed from [6].

2.3 Emission Rates and Control Efficiencies

Except for the handling of dusty material, particulate emissions nearly always originate during chemical reactions occurring in large vessels that contain heated solids and liquids. Dust suspensions in hot gases flow from the vessel through a system of conduits terminating in an open stack. If no mechanical devices are placed along the gas stream to retrieve particles, resulting discharges to the atmosphere are known as "uncontrolled" or "raw" emissions. The "raw emission rate" is usually expressed as weight of particulate per unit-weight of heated material in the vessel. Emissions arrested by collecting devices are called "controlled emissions." Those that are not collected by such devices are "residual emissions." Controlled and residual emissions are measured in the same units as the raw emission rate. The percentage (by weight) of total emitted dust collected by a device is the "collection efficiency" of the device.

Seldom do all plants in a given industrial or power-generation process trap their escaping particulates with the same collection efficiency. In fact, some plants may exercise no pollution control (use no collecting devices) at all. The "application factor" of control in a polluting process is the percentage of plants employing some control. For process j, let

N_j = weighted average collection efficiency
M_j = application factor
P_j = raw emission rate
A_j = controlled emission rate
R_j = residual emission rate

Then, define

$$P_j = A_j + R_j \tag{1}$$
$$A_j = N_j M_j P_j \tag{2}$$
$$R_j = (1 - N_j M_j) P_j \tag{3}$$

Whereas the input-output "production function" is linear and homogeneous of degree one, the production function of a partic-

ulate collection system is nonlinear in its inputs. Why is this so?
To begin with, all dust collectors place a series of obstructions
in the way of moving particles. An obstructing "unit" in a collection system may be a textile fiber, a jet of water, or a mag-
netic field. (Each unit is distinguished from its neighbor by the
time at which the advancing dust cloud touches it. Thus we
may speak of an infinite number of obstructing units in a
physically continuous medium like a precipitator plate or fabric
filter bag.)

Collection efficiency may be increased by (1) upgrading the
effectiveness of each obstructing unit and/or (2) increasing the
number of units. The first alternative requires technical improvement and involves more than mere adjustment of the parameters
of a static production function. Only by using more and larger
collectors does one improve collecting power without a change
of technique. Adding the nth obstructing unit to a system affects
only those dust particles not trapped by the first through $(n-1)^{th}$
units. For N obstructing units, each capable of removing a fraction u of the dust loading of oncoming gases, total collection
efficiency (in percent) is

$$100 \left[1 - (1-u)^N\right]$$

For example, a system using two 45 percent efficient units will
collect not 90 percent of all particles (as in a linear system) but
$100[.45 + .45 (1-.45)]$ or 69.75 percent. Measured in dollars
per unit of dust collected per minute, the cost of abatement
rises at an increasing rate with percentage collection efficiency.

2.4 Particulate Abatement as an Input-Output Activity

This nonlinear aspect of abatement technology makes the use of
an input-output model for particulate collection a hazardous
business. Since average collection cost is proportional to the collection rate, all collector-cost or input coefficients are valid for

only a narrow range of collecting-device efficiency. In part, the amount of abatement service which each sector will pay for depends on the cost implicit in abatement technology. But a sector may well have a choice among alternative collection efficiencies. Accordingly, a column vector of abatement activity in the input-output pollution model really pertains to specific levels (or narrow ranges) of both collection efficiency and collection technology.

What value or range of collection efficiency should be built into the input-output pollution model? To give an answer, we need to know the intended use of this highly flexible model. To measure actual present-day flows of resources into the abatement sector, one would simply use the average collection efficiency of existing devices (aggregate controlled emissions as a percentage of aggregate raw emissions). To arrive at an abatement-input coefficient for an individual industry, one would need to compute the ratio of *actual dollar flows to the abatement sector* to *the gross output of the polluting sector.* If only plant-by-plant emissions and efficiencies were known, cost estimates could be read from graphs similar to those of Figure 1.

To provide coefficients for static and dynamic interindustry impact analyses of recent federal air pollution regulations, we must predict the collection efficiency levels that polluting sectors will employ to satisfy these new limits on their emission rates. The methodology used in making these predictions is presented in section 4. Three measures of collection efficiency must be distinguished.

(1)   "Actual" efficiency is the average efficiency level of gas cleaning currently maintained by a polluting industry (aggregate controlled emissions as a percentage of aggregate raw emissions).

(2)   "Required" collection efficiency is the efficiency level that reduces a sector's average raw emission rate to the emission rate imposed by legal standards.

(3)   The "target" efficiency level is a collection efficiency

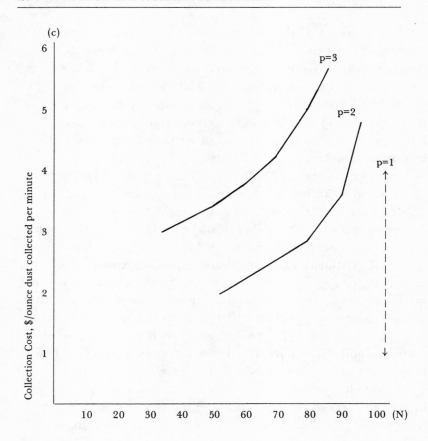

Figure 1. *Percentage Collection Efficiency*

level that enables a sector to *at least* meet legal standards. The target efficiency may be greater than the required level if generally available devices overshoot government standards.

2.5 Particulate Collection Technologies

In our calculations, the difference between required and target efficiency levels was never very large, but was usually substantial enough to warrant the distinction. The origin of the discrepancy is best explained in a discussion of the three types of high-efficiency particulate collection machinery:

(1) electrostatic precipitators
(2) fabric filters
(3) wet scrubbers.

Electrostatic precipitators currently find very wide particulate control applications. Inside a cube-shaped housing is a set of metal plates. Each plate is the length and height of the machine and faces at right angles to the oncoming gas flow. It stands about six to 12 inches from its neighbor. Precipitation involves charging the plates electrostatically so that oppositely charged dust particles will be attracted to them. At regular intervals a mechanical rapping device is activated to dislodge clinging dust from the plates. The dust falls into collection hoppers for eventual disposal. A precipitator can give almost any desired collection efficiency if built to the correct size relative to the volume, velocity, and dust content (or "loading") of the gas to be treated. (The crucial parameter of precipitator size is the total area of the collecting plate.) Therefore, required and target efficiency levels will equal each other for all precipitator applications. Precipitator materials can be treated to resist very hot and corrosive gases, but the temperature and surface texture of particulate matter affects its electrical resistivity and thus the ease of precipitation. If N is precipitator collection efficiency, Q is total plate surface area in square feet, W is the precipitation rate in feet per minute, V is the volumetric gas flow rate in actual cubic feet per minute (acfm), and e is the basis of natural logarithms, then

$$N = 100 \left[1 - e^{-W(Q/V)}\right] \tag{4}$$

This is the Deutsch-Anderson equation, a basic relationship in precipitation engineering. Its importance will become apparent in the discussion of precipitation cost estimation in Section 5.

A fabric filtration unit or "baghouse" is best suited to very small particles (two microns in diameter or less). Baghouses cannot tolerate as much heat and corrosion as precipitators. Each machine consists of a number of cylindrical or envelope-shaped filters of woven or felted textile material placed in rows inside a metal housing. Contaminated gases flow into the bags much as fingers enter a glove. Particulate matter collects on the inner surface of each filter and falls or is shaken off into a collecting hopper. A recent NAPCA study explains that collection efficiency is determined by the accumulated dust layer inside each filter.

> As dust accumulates [inside a new filter], efficiency rises. By the time the deposit amounts to 2 to 3 grams of dust per square meter ... the collection efficiency usually exceeds 90%. This happens in about one minute at ... a common industrial gas loading. Overall efficiency then continues to increase with dust and accumulation, and will generally exceed 99% ... after an hour or less at common industrial dust concentrations. After a period of cyclic filtration and cleaning ranging from a few hours to a few days, the residual deposit will stabilize and thereafter efficiency will remain greater than 99%. During usual operating conditions fabric filter overall weight collection efficiency will exceed 99.9%. [5]

If a baghouse is at all suited to its application, it will render at least 99.9 percent collection efficiency. Thus if a polluting industry were confronted with a required collection efficiency of 98 percent, it could purchase a precipitator designed especially for that requirement or one that gives 99.9 percent efficiency. Polluting sectors that use baghouses to comply with Clean Air Act standards will have target efficiencies of 99.9 percent regardless of required efficiency. This difference between required and target efficiencies is small.

A third high-efficiency dust collector is the wet scrubber. A jet of scrubbing liquor (usually water) obstructs moving particles. The dust is trapped in the liquor and directed toward a disposal unit. Scrubber efficiency depends mostly upon "contacting power," the energy spent in obstructing particles measured in horsepower per acfm (actual cubic feet per minute). Scrubber efficiency is less readily varied by the designer than precipitator efficiency and hence target and required efficiencies of scrubbers diverge.

3.0 DERIVATION OF THE COEFFICIENTS

In this section we express each pollution and abatement sector coefficient in terms of parameters whose values appear in the pollution control literature. This provides a framework for the more detailed discussions of data and methodology presented in sections 4–6.

3.1 Emissions Coefficients

Three sets of emission coefficients are estimated. They measure a polluter's rate of generation of raw, controlled, and residual emissions. Each coefficient is measured in tons of particulate matter per dollar of gross output of the polluting sector. Each raw emission coefficient is the sum of the abated and residual emission coefficients for the sector in question. Each controlled emission coefficient is the product of the raw emission coefficient and the sector's target collection efficiency level. Emission coefficients and target efficiencies were estimated for polluting processes within each input-output sector and then aggregated into sectoral coefficients. Thus each sectoral coefficient is really a weighted average of process-level coefficients.

Sectoral raw emission coefficients for a particular base year were obtained by summing the tonnages of raw emissions for all the sector's polluting processes and dividing that sum by gross domestic output. For example, the raw effluent tonnages of five high-pollution steel processes plus the raw dust output of a generalized "other steel processes" category were summed to make the numerator of the steel sector's raw emission coefficient. (Process raw emission tonnages were sometimes obtained directly and in other cases calculated as the product of raw pollution per tons of process output—an engineering estimate—and process output tonnage.) Similarly, controlled and residual emissions coefficients were estimated by dividing estimates of total controlled and residual emissions by sectoral output. The emission level for each process was multiplied by target collection efficiency to estimate its controlled tonnage. Raw tonnage less controlled tonnage yielded residual tonnage.

Interpretation of Clean Air Act standards permitted a division of raw emissions into controlled and residual tonnages. In all cases, we assumed that governmental regulation would bring application factors of 100 percent. For all processes with "good" data (as defined in section 2.2), it was possible to estimate allowable residual emission rates under Clean Air Act regulations. The methodology of these estimates is explained in section 4. Subtracting allowable residual emission rates from raw emission rates gave the controlled emission rates required under the law. These controlled rates were in turn divided by the raw rates to yield required collection efficiencies. Finally, comparing required efficiencies with those of high-efficiency collection devices (see section 4.3) enabled us to estimate a target efficiency level for each "good-data" process. Since all these target efficiencies hovered about a rough mean of 99 percent, we took 99 percent as our target collection efficiency for all polluting processes where good data were unavailable. Thus, for all "poor-data" sectors, the target efficiency prediction yielded the required controlled (and thus residual) emission rate. The procedure for "good-data" sectors was the reverse of that for "poor-data" sectors.

3.2 Abatement Coefficients

Abatement coefficients measure purchases of abatement services per dollar of gross output of the polluting sector. Each sector buys enough abatement services to attain its target collection efficiency under the Clean Air Act. For the "good-data" processes, annual abatement costs were computed as the product of three factors: cost per acfm, acfm per ton of output, and annual output. To aggregate all processes in a given sector, it was necessary to assign value weights to individual polluting processes. Where intraplant operations were singled out as separate processes in the literature (e.g., steel sintering), it was difficult to impute values to their outputs. In some cases, arbitrary assumptions were required to estimate weights for such processes. Weighting was less of a problem where direct estimates of total abatement costs for particular processes were available. An average value for all "good-data" processes of dollar flow per ton of raw emission was applied to the raw emission tonnage of each "poor-data" process.

3.3 Input Coefficients for Abatement Sectors

We obtained the abatement sector's input structure from cost breakdowns for specific control devices discussed in section 6. To estimate a column vector for the abatement industry we computed the percentage distribution of abatement costs by industry of origin for precipitators, fabric filters, and wet scrubbers. These percentage distributions had to be weighted by estimates of the relative volume of expenditures on each type of collector. This is equivalent to adding predicted outlays by all "good-data" sectors in each cost category and expressing this single new aggregated cost breakdown in percentage terms. Distinct input structures were derived for the current and the capital accounts of the abatement sector. Current input coefficients are simply the proportions of components to total cost. Capital coefficients are capital-cost proportions multiplied by the abatement sector's average capital-output ratio. Replacement coefficients multiply the capital coefficient vector by the inverse of expected mechanical service life. We used the commonly cited value of fifteen years for average service life. The capital-output ratio is the total required capital stock in abatement divided by the total value of abatement services. In the process of computing total required capital stocks, we obtained values of capital stocks for abatement per dollar of polluting sector output for each input-output sector. These estimates can be used to assign capital stocks in the dummy abatement sector to the clean-up requirements of individual polluting industries.

4.0 EMISSION COEFFICIENTS: DATA AND METHOD

Raw emission rates (tons of particulates per ton of process output) are truly engineering parameters based upon chemical relationships. They are not subject to year-to-year fluctuations; only an alteration of process chemistry can change them. But when we combine the raw emission rates of several processes to obtain sectoral raw emission coefficients (tons of particulates per dollar of sectoral output), the value of that coefficient depends on the relative importance of the individual processes in the sector. For example, the steel sector's raw emission coefficient has changed in recent years as open hearth furnaces have been replaced by basic oxygen and electric arc furnaces. For this reason, any input-output pollution model is identified with a particular year or period of a few years. Our process output and input-output flow data pertain to 1970. But the basic process information can be reweighted for other years.

Studies by the Midwest Research Institute (MRI) [6] and the Southern Research Institute (SRI) [7] provided a large part of our data and all our raw emission rates. The MRI study of particulate systems was used more extensively than the SRI electrostatic precipitation monograph, although in general these authorities were in close agreement. The process raw emission rates and target efficiency levels shown in Table 2 may be used to develop sectoral emission coefficients for other base years.

Table 2. *Process Raw Emission Rates, Target Efficiency Levels, and Tonnage Outputs of Selected Products (1970)*

Process	Emission Rate (lbs./ton)	Output (000 tons)	Target Efficiency (%)
Coal combustion			
coal-burning electric utilities	135.5	329,945*	98.4
other coal combustion	135.5	105,923*	99.0
Steelmaking			
blast furnaces	107.5	91,816	99.9
basic oxygen furnaces	41.6	63,330	99.4
open hearth furnaces	19.1	48,022	92.0
sintering	20.0	45,608	97.8
electric arc furnaces	17.7	20,162	99.9
other steel processes	6.3	594,995	99.0
Cement			
wet process	167.0	47,544	99.9
dry process	167.0	29,656	99.6
other cement processes	42.5	77,963	99.0
Crushed stone, sand, gravel	7.2	1,690,969	99.0
Asphalt			
paving materials	40.0	287,900	99.9
other asphalt processes	6.4	8,125	99.0
Forest products			
pulp mill recovery boilers	150.0	29,579	99.3
other pulp and paper processes	65.0	30,422	99.0
other forest products	10.0	41,400	99.0
Nonferrous metals	24.0	288,135	99.0
Agricultural processes	25.0	207,920	99.0
Fertilizer and phosphate rock	23.0	173,304	99.0
Lime			
rotary kilns	180.0	21,200	99.9
other lime processes	16.0	54,625	99.0
Clay products	80.0	31,825	99.0
Ferroalloys	268.0	1,875	99.0
Iron foundries	13.4	31,652	99.0
Acids	6.8	41,468	99.0
Coal cleaning	2.6	73,843	99.0
Carbon black	132.0	1,657	99.0
Petroleum	76.0	1,210	99.0

*Output figure denotes coal usage.

Sources: [6, 7].

Note that the "fuel combustion" entry of Table 1 appears as "coal combustion" in Table 2. Fuel oil and natural gas combustion have particulate emission rates too low to warrant controls under the 1970 Clean Air Act standards.

MRI also published estimates of output (in physical units), average collection efficiency and application factor for the year 1968. These production figures were updated to 1970 on the basis of industrial output data from the Census Bureau [8]. Where MRI industrial process categories did not strictly match those of the Census, we employed a production index based upon Census figures for 1968 and 1970. Extrapolation of Census data was necessary when a 1970 observation was not available. A detail of electric utility output by generator fuel usage in [4] enables us to apply the coal-combustion emissions rate to output data for coal-burning utilities. Estimates of the 1970 output of all processes appear in Table 2. To estimate controlled and residual emissions rates we need to know target efficiency and for that, required efficiency. Federal strategy for implementing the Clean Air Act is, however, to develop national air quality guidelines. Within these national guidelines, each state sets its own standards for residual emissions [3]. Federal emission ceilings have been published for only two major particulate sources, electric power and cement plants [2]. Only time will reveal the average required efficiency levels that grow out of continued federal and state regulatory activities. Implementation of the Act is by no means complete, and may not be so until 1980.

Since we did not know required efficiency levels for most sectors, we estimated allowable emission rates from generalized formulae for "attainable" particulate residual emission rates for process industries. These formulae are attached to an Environmental Protection Agency (EPA) plan for implementation of the Clean Air Act [2]. If H_r is the "attainable" residual emission rate in pounds per *hour* (rather than pounds per ton of material processed), and H_w represents the "process weight rate" in tons of material processed per hour, then

$$H_r = 3.59\ H_w{}^{.62} \qquad H_w \leqslant 30 \tag{5}$$

$$H_r = 17.31\ H_w{}^{.16} \qquad H_w > 30 \tag{6}$$

Process weight rates were computed for "good-data" sectors from information in [6]. Note that these formulae do not reduce to a single attainable residual emission rate in pounds per ton of material processed. According to the formulae, residual emissions in pounds per ton should decrease as tons-processed-per-hour increases. If T is the attainable rate in pounds per ton of material processed, the equations (5) and (6) become

$$T = 3.59\ H_w{}^{-.38} \qquad H_w \leqslant 30 \tag{5a}$$

$$T = 17.31\ H_w{}^{-.84} \qquad H_w > 30 \tag{6a}$$

In other words, the residual emission rate (pounds per ton) varies inversely with the process weight rate (tons per hour). The allowable residual emission rate for "poor-data" processes was set at one percent of the raw rate in view of the proximity of this figure to computed allowable rates.

The MRI and SRI studies suggested the types of high-efficiency collection equipment most likely to be used to meet the standards. From this information, target efficiency levels for precipitator and fabric filter applications followed immediately. Target efficiency levels were set equal to required levels for all precipitator applications and were set to 99.9 percent for all fabric filter applications. There is one exception to this rule. For the obsolete open hearth process of steelmaking, target efficiency was set at the average efficiency of existing collectors (92 percent). In other words, we assumed that the EPA will require some control (the average existing collector being the standard) on all open hearth furnaces, but will not expect expensive high-efficiency equipment to be installed on furnaces now nearing retirement. The application factor of low efficiency devices on open hearth furnaces is only about 40 percent. We assumed that there would be substantial steel industry investment in additional lower-efficiency collectors for its open hearth furnaces. Finally, a discussion of particulate control in the MRI study indicates that high-efficiency wet scrubbers in steel blast furnaces will meet the required efficiency level of 99.9 percent. Target efficiency estimates appear in Table 2.

5.0 ABATEMENT COEFFICIENTS: DATA AND METHOD

Abatement coefficients are the ratio of the total annual abatement expenditure of each polluting input-output sector (an aggregate of one or more polluting processes) to the sector's gross output. Since 1970 sectoral gross outputs appeared in input-output literature (see section 6), and 1970 process tonnages had been estimated (Table 2), it remained to estimate:

(1)   acfm per ton of process output
(2)   current costs of abatement per acfm.

For "good-data" processes, the product of process tonnage and the two factors noted above was the predicted abatement outlay. As noted in section 3, we summed outlays made by all "good-data" processes and divided these figures by their total raw emission tonnage to arrive at rough averages of current-account abatement cost per ton of raw emission. This parameter is reasonably stable across processes. Total abatement costs for "poor-data" processes were then estimated by multiplying their total raw emissions by these average unit abatement costs. The sum of estimated outlays of all processes is the total estimated expenditure of each input-output sector.

5.1 Sources of acfm per Ton of Output

The magnitude of a gas-cleaning job depends on two factors: the rate of gas expulsion in acfm per ton per year of output, and the physical and chemical nature of the gas. Expulsion rates for many precipitator applications were computed from [6] and a few from [7]. The former source listed values of standard cubic feet of gas generated per ton of process output as well as vessel temperature and pressure. Actual cubic feet per minute per ton per year of process output was computed by adjusting standard cubic feet per ton of output for actual temperature and pressure and then dividing by the number of minutes in a working year (420,000–480,000, depending upon the process).

5.2 Sources of Collector Cost per acfm

The best engineering data for current operating and maintenance costs of particulate control (exclusive of capital charges) came from [10]. But even these estimates were in the form of generalized formulae for collectors of various efficiency ranges. Where the cost formulae permitted choices of parameter values to reflect both the difficulty of the gas-cleaning job and the desired efficiency level, we chose average values of the former and the highest values of the latter.

Table 3a contains estimates of collector costs on current account per ton of process output for all polluting processes. Capital charges are not included at this point. These numbers are analogous to the process emission rates of Table 2. To reiterate, once capital charges have been added on, the coefficients in Table 3a can be multiplied by process output tonnages for any base year to estimate outlays for particulate abatement for each process. Process outlays are summed by input-output sector and divided by sectoral gross outputs to give a set of abatement coefficients for input-output sectors. All figures are in 1958 dollars.

Table 3a. *Current Operating and Maintenance Coefficients for Particulate Abatement (Excluding Capital Charges) for U.S. Industrial Processes (in 1958 dollars per ton of output)*

Process	Coefficient
Coal combustion	
coal-burning electric utilities*	.057
other coal combustion*	.057
Steelmaking	
blast furnaces	.263
basic oxygen furnaces	.147
open hearth furnaces	.265
sintering	.040
electric arc furnaces	.149
other steel processes	.003
Cement	
wet process	.114
dry process	.114
other cement processes	.021
Crushed stone, sand, gravel	.004
Asphalt	
paving materials	.004
other asphalt processes	.003
Forest products	
pulp mill recovery boilers	.119
other pulp mill processes	.033
other forest products	.005
Nonferrous metals	.012
Agricultural processes	.013
Fertilizer and phosphate rock	.012
Lime	
rotary kilns	.052
other lime processes	.008
Clay products	.040
Ferroalloys	.134
Iron foundries	.007
Acids	.003
Coal cleaning	.001
Carbon black	.066
Petroleum	.038

*Cost figures in dollars per ton of coal burned.

6.0 INPUT STRUCTURE OF THE ABATEMENT SECTOR: DATA AND METHOD

The coefficients of the abatement sector itemize the bill of materials, services, and labor on capital or current account for the installation or operation of a pollution control device. Each item is classified by the input-output sector of its origin. Labor cost and capital charges are assigned to the value-added row. Since all valuations are in producers' prices, the trade row contains the trade margin on a dollar's worth of abatement purchases.

As a "dummy" activity, the abatement sector is an accounting convenience in the input-output pollution model that permits us to segregate abatement spending from the production of normal economic goods.* Abatement activity usually involves the use of machinery to collect pollutants. Ordinarily, abatement equipment is purchased and operated by producers of goods who emit pollution as a by-product of their operations. In the present input-output accounting we treat the abatement sector as an autonomous industry that owns and operates control machinery and sells the service of pollution abatement to polluting industries.

6.1 The Cost Breakdowns

Our only source for particulate collector installation and operating cost breakdowns was, fortunately, a very good one. Investigators for Battelle Memorial Institute studied the estimate files of the Swindell-Dressler Company, an abatement device contractor. The information yielded estimated cost breakdowns for all major collection equipment in several steel industry applications.

---

*Although many firms are engaged in the sale of pollution abatement equipment, they commonly subcontract for all but engineering tasks and the top-level supervision of the installation process. They serve primarily as technical advisers and purchasing agents for their customers. Thus the dummy sector characterization is not entirely unrealistic.

Specific dollar amounts in the Battelle cost breakdowns thus apply to a limited range of collection efficiencies and applications. In percentage terms, however, these itemized lists exhibit marked similarities among different devices and applications. Accordingly, it seemed reasonable to apply a single set of abatement sector column proportions to all our specific particulate applications and efficiencies. Sectoral differences in application and efficiency are still, of course, registered in differences in coefficients along the abatement sector row. If, for example, our target collection efficiency levels were closer to 50 percent than to 99 percent, coefficients along the abatement row (dollars of abatement purchases per dollar of polluting industry gross production) would be considerably lower, while proportions in the abatement column might remain constant.

The actual computation of the generalized abatement cost breakdown for current- and capital-account items was a two-stage procedure. (The conversion of these projections into input-output column vectors and the computation of trade and transportation margins is discussed in the next section.) First, we examined the Battelle cost breakdowns in percentage terms to determine an average breakdown for each of the three principal types of collection devices. In the absence of more specific information, weighting was based on judgment. The second step was to obtain generalized percentage-cost breakdowns representing all three collectors. To do this, we weighted the three device-specific average breakdowns. For each type of device, the weight used was the predicted total outlay by all "good-data" processes expected to use the device to meet federal standards. Percentage cost breakdowns were estimated separately for current- and capital-account items. The proportions should be regarded as rough rules of thumb rather than precise estimates.

6.2 The Abatement Sector Capital-Output Ratio

An estimate of the average capital-output ratio for the abatement sector was required to convert the percentage distribution of capital costs into a column of abatement sector capital coefficients (value of capital stock held per dollar of current output) and to estimate replacement coefficients. The capital-output ratio is the quotient of total capital stock in abatement and corresponding total current outlays. Abatement capital is counted as "held" by the abatement sector, and its services are reflected in abatement coefficients.

Estimating total capital stock in abatement was a fairly complex procedure. Most sources quote collector cost net of transportation and installation charges (i.e., f.o.b.). Estimates of collector cost per acfm for various particulate applications were found in four sources. First, all steel process abatement costs were taken from a cost analysis of air pollution control in the steel industry [5]. Secondly, MRI presented estimates of control cost per acfm for several nonsteel applications. Thirdly, for all nonsteel, nonprecipitator allocations not covered by MRI, a set of cost curves for various control devices in [10] sufficed. These device-cost curves do not take the chemistry of particular process into account and should be less accurate than process-specific information. They were used only as a last resort.

The source of nonsteel precipitator costs was a slide chart issued by the Belco Pollution Control Corporation, a major precipitator designer and contractor. The chart relates f.o.b. shop precipitator price to total collecting plate area. Combining the slide chart with a transformation of the Deutsch-Anderson equation (above, p. 235), we expressed precipitator price as a function of important engineering and price variables, including an efficiency factor. This permitted us to equate required and target efficiency levels for precipitator applications. Our first step in arriving at the price formulation was to transform the Deutsch-Anderson equation into an expression for Q or total collecting plate area.

$$Q = (V/W) \ln (100/100-N) \tag{7}$$

Next a regression line was fitted to points read off the Belco slide chart.* If C stands for f.o.b. shop price in 1968 dollars, then

$$C = 30,620 + 1.88Q \tag{8}$$

Replacing Q in equation (8) with the right-hand term in equation (7) gave the f.o.b. cost equation. The SRI paper provided values of W, the precipitation rate parameter. The intercept term in (8) incorporates an assumption about the number of precipitators installed at each application site: it was assumed to be one. Inserting SRI's value of W, target efficiency as N, and average process acfm per ton of yearly goods output for V, we estimated cost in terms of the annual output rate. Then cost was divided by acfm per ton to yield an estimate of f.o.b. cost per acfm per ton of yearly goods output. The average ratio of total capital cost to f.o.b. collector cost came directly from our percentage breakdown of capital costs. The ratio was simply the inverse of the "collector" component of the capital breakdown. Its value was about 6.4. Inserting the values of acfm per ton of process output used to calculate abatement coefficients led to estimates of f.o.b. collector and total capital cost per ton of process output. (See Table 3b.)

7.0 THE COEFFICIENTS IN INPUT-OUTPUT FORMAT

Thus far we have developed forecasts of flows of materials, services, and labor that will be required for industrial abatement of particulates under the Clean Air Act. But our object is to develop coefficients for the special dummy sectors so that the input-output pollution model can be put to work. Two steps remain to achieve this goal. First, we must assign our 29 particulate-polluting industrial processes, and each item in our collecting-device material breakdowns, to input-output sectors. Second, the

---

*This linear expression gave a nearly perfect fit.

Table 3b. *F.O.B. Collector and Total Capital Costs of Abatement for U.S. Industrial Processes (1958 dollars per ton of output)*

Process	F.O.B. Collector	Total Capital
Coal combustion		
coal-burning electric utilities*	.42	3.77
other coal combustion*	.42	3.77
Steelmaking		
blast furnaces	1.10	6.46
basic oxygen furnaces	.21	1.03
open hearth furnaces	.51	2.34
sintering	.18	.46
electric arc furnaces	.22	1.32
other steel processes	.02	.16
Cement		
wet process	1.03	4.41
dry process	.45	2.26
other cement processes	.17	1.06
Crushed stone, sand, gravel	.03	.18
Asphalt		
paving materials	.02	.08
other asphalt processes	.02	.16
Forest products		
pulp mill recovery boilers	1.35	7.33
other pulp mill processes	.25	1.63
other forest products	.04	.25
Nonferrous metals	.09	.60
Agricultural processes	.10	.63
Fertilizer and phosphate rock	.09	.58
Lime		
rotary kilns	.49	2.52
other lime processes	.06	.40
Clay products	.31	2.00
Ferroalloys	1.05	6.70
Iron foundries	.05	.34
Acids	.03	.17
Coal cleaning	.01	.07
Carbon black	.52	3.30
Petroleum	.30	1.90

*Cost figures in dollars per ton of coal burned.

terms in each vector of flows must be divided by the gross output of the appropriate sector to yield coefficients. Current, replacement, and capital coefficients for goods-producing sectors are already available. It remains to compute coefficients for the dummy sectors.

7.1 The Classification Procedure

We adopted the 83-order input-output aggregation scheme of the Bureau of Economic Analysis (BEA), in view of its widespread usage in current economic research. The base year for price deflation was 1958. Competitive imports were removed from the system to create a domestic base table. Table 4 shows the alignment of input-output sectors with polluting processes. Table 5 shows the industry of origin to which various inputs to abatement were assigned. Since labor is a component of value added, it is classified as such under sector "VA."

The entry "other coal combustion" in Tables 2 and 3 subsumes all industrial users of coal other than the principal user, electric power generation. The distribution of flows associated with "other coal combustion" was computed from data supplied by the Bureau of Mines [11]. The Bureau, in turn, assigns about 17 percent of industrial coal consumption to an "all other" category. The remaining fraction was distributed in proportion to the "coal mining" row of the BEA table of input-output flows for 1963.

7.2 Computation of the Coefficients

A vector of 1970 gross domestic outputs was computed by multiplying the Leontief inverse of 1970 technical coefficients from [12] by a vector of 1970 final demand from [13]. The coefficient table was converted to domestic base by removing competitive

Table 4. *Classification of Particulate-polluting Industrial Processes According to the 83-order BEA Aggregation Scheme*

Sector Number	Sector	Polluting Process
1	livestock	coal combustion
2	other agricultural prods.	agricultural processes
5	iron mining	coal combustion
7	coal mining	coal cleaning
9	stone and clay mining	coal combustion
14	food & kindred prods.	coal combustion
16	fabrics, yarn & thread	coal combustion
20	lumber and wood prods.	other forest products
24	paper	pulp mill recovery boilers
		other pulp and paper processes
		coal combustion
27	chemicals	fertilizer and phosphate rock
		carbon black
		acids
		coal combustion
28	plastics	coal combustion
29	drugs and cleaning prepar.	coal combustion
31	petroleum and products	asphalt paving materials
		other asphalt processes
		petroleum
32	rubber and misc. plastics	coal combustion
36	stone and clay products	wet process cement
		dry process cement
		lime rotary kilns
		other lime processes
		crushed stone, sand, gravel
		clay
		coal combustion
37		(all steel processes)
		ferroalloys
		iron foundries
		coal combustion
38	nonferrous metals	nonferrous metals
		coal combustion
59	motor vehicles	coal combustion
65	transportation	coal combustion
68	utilities	electric power
71	real estate	coal combustion
77	medical and nonprofit	coal combustion
79	state and local gov't.	coal combustion

Table 5. *Classification of Suppliers of Abatement Sector According to the 83-order BEA Aggregation Scheme*

Sector Number	Sector	Supplier of Abatement
	CAPITAL ACCOUNT	
11	new construction	foundation
40	fabricated structural metal	water treatment, piping
		ducts, stack
		structural supports
49	general industrial machinery	collector
		fan, motor
55	electric and wiring equipment	electric and wiring equipment
62	scientific and control instr.	control instruments
65	transportation	transportation
69	trade	trade
73	business services	engineering
VA	value added	labor
	CURRENT ACCOUNT	
12	maintenance construction	maintenance supplies
68	utilities	electric power
		water
VA	value added	labor

imports, thereby yielding a gross output vector that counts domestic production only.

7.2.1 Final Coefficients: The Pollution Sector Row

Table 6a contains estimated 1970 flows of raw particulate emissions tonnage by 83-order sector according to the quality of available data. The emission levels of coal-burning sectors other than utilities are presented separately because they are based on the same coal combustion emission rate. Target efficiency levels of particulate collection were multiplied by the raw emission tonnages of Table 6a to estimate the residual tonnage in Table 6b.

Table 6a. *Raw Particulate Emission Tonnage of Industrial Processes in 1970 by Input-Output Sector and Quality of Data (000 tons)*

I-O No.	"Good Data" Sectors	"Poor Data" Sectors		Total
		Coal-burning	Other	
1		29		29
2			2604	2604
5		19		19
7			99	99
9		19		19
14		323		323
16		58		58
20			129	129
24	2214	567	1068	3849
27		818	2244	3062
28		106		106
29		19		19
31	5758		69	5827
32		38		38
36	8354	495	9381	18230
37	7343	2395	2326	12064
38		1384	3464	4848
59		68		68
65		38		38
68	22354			22354
71		48		48
77		48		48
78		319		319
79		377		377
Total	46023	7168	21384	74545

A generous margin of error surrounds most of the entries at this early stage in the enforcement of the Clean Air Act. Controlled emissions are not listed separately; they are simply the difference

Table 6b. *Residual Particulate Emission Tonnage of Industrial Processes in 1970 by Input-Output Sector and Quality of Data (000 tons)*

I-O No.	"Good Data" Sectors	"Poor Data" Sectors		Total
		Coal-burning	Other	
1		.46		.46
2			26.04	26.04
5		.30		.30
7			.99	.99
9		.30		.30
14		5.17		5.17
16		.93		.93
20			1.29	1.29
24	15.53	9.07	10.68	35.28
27		13.09	22.44	35.53
28		1.70		1.70
29		.30		.30
31	5.58		.69	6.27
32		.61		.61
36	15.79	7.92	93.81	117.52
37	59.74	38.32	23.26	121.32
38		22.14	34.64	56.78
59		1.09		1.09
65		.61		.61
68	357.66			357.66
71		.77		.77
77		.77		.77
78		5.10		5.10
79		6.03		6.03
Total	454.30	114.68	213.84	782.82

between raw and residual emissions. According to our calculations, federal restrictions on particulate pollution can reduce per-capita emissions in the United States from about 750 pounds

Table 6c. *Raw and Residual Coefficients Along the Pollution Sector Row (tons per billion 1958 dollars)*

I-O No.	Raw Coefficients	Residual Coefficients	I-O No.	Raw Coefficients	Residual Coefficients
1	908	14	31	198373	213
2	92611	926	32	2365	38
5	8999	142	36	1698737	10951
7	25997	260	37	367023	3691
9	8557	135	38	237483	2781
14	3660	59	59	1181	19
16	3056	49	65	690	11
20	11260	113	68	520974	8335
24	204351	1873	71	436	7
27	112371	1304	77	1188	19
28	8644	139	78	47729	763
29	1315	21	79	40768	652

to approximately eight pounds annually. Coefficients along the pollution sector row appear in Table 6c.

7.2.2 Final Coefficients: The Abatement Sector Row

Estimated purchases by polluting industries of the services of the dummy abatement sector are listed in Table 7a and the corresponding abatement coefficients are given in Table 7b.

7.2.3 Final Coefficients: The Abatement Sector Column

Since input-output coefficients for goods in the 1970 table are in producers' prices, we computed transportation and trade coefficients for the abatement sector. Purchases from construction and utilities on current account have zero margins. Transport and trade coefficients are included in the capital

Table 7a. *Dollar Flows per Annum on Current Account for Particulate Abatement Sufficient to meet Clean Air Standards, by Input-Output Sector and Quality of Data, 1970 (thousands of 1958 dollars)*

I–O No.	"Good Data" Sectors	"Poor Data" Sectors		Total
		Coal-burning	Other	
1		24		24
2			2604	2604
5		16		16
7			99	99
9		16		16
14		271		271
16		49		49
20			129	129
24	3513	476	1068	5057
27		686	2244	2930
28		89		89
29		16		16
31	1237		69	1306
32		32		32
36	9904	415	9381	19700
37	61741	2009	2326	66076
38		1161	3464	4625
59		57		57
65		32		32
68	18752			18752
71		40		40
77		40		40
78		268		268
79		316		316
Total	95147	6013	21384	122545

Table 7b. *Coefficients along the Abatement Sector Row*
(cents per 1958 dollars)

I-O No.	Coefficient	I-O No.	Coefficient
1	.0001	31	.0090
2	.0093	32	.0002
5	.0008	36	.1836
7	.0026	37	.2010
9	.0007	38	.0227
14	.0003	59	.0001
16	.0003	65	.0001
20	.0011	68	.0437
24	.0268	71	
27	.0108	77	.0001
28	.0007	78	.0040
29	.0001	79	.0034

column. They were calculated as the sum of trade margins [9] on all items in a typical dollar's worth of abatement capital. The composition of this typical dollar was the generalized collector-cost breakdown (see section 6) categorized according to the BEA's 83-order system. The abatement sector's capital coefficient column vector was the product of the estimated capital-output ratio (section 6.2) and the percentage distribution of capital cost by 83-order sector of origin. The current replacement and capital coefficients appear in Table 8.

One by-product of the estimating process was a capital-account "cleaning bill" for each polluting sector (Table 9). Dividing a polluting sector's capital "bill" by its gross output yields an abatement capital coefficient for each sector. They can be used to estimate the impact of changes in the level and composition of final demand on capital costs for abating the pollution of each sector.

Table 8. *Abatement Sector Column Coefficients for Current and Capital Matrices (cents per 1958 dollar of current-account abatement purchases)*

I–O No.	Capital Coefficient	Replacement Coefficient	Current Coefficient
11	58.10	3.87	
12			2.74
40	849.45	56.63	
49	529.09	35.27	
55	202.75	13.52	
62	66.97	4.46	
65	31.79	2.12	
68			5.18
69	127.57	8.50	
73	391.66	26.11	
VA	1088.44	72.56	92.08

8.0 CONCLUSION

The increasing demand for environmental intelligence may eventually spawn a comprehensive system of "pollution accounts" incorporating surveys of emission and control for all major pollutants from all major sources. This study suggests an *ad interim* research strategy for constructing a working input-output pollution model. A few good observations of air particulate emissions and control activities have sufficed for preliminary projections of the necessary coefficients. Briefly, each type of coefficient was separated into component variables for which data were fairly readily obtainable. As one might expect, the single serious limitation to obtaining accurate environmental coefficients is the relative scarcity of reliable control cost data. Expanding public efforts to check the pollution crisis will require substantially increasing investments in high-efficiency control facilities. As

Table 9. *Capital Stocks of Particulate Abatement Equipment Sufficient to Meet Clean Air Act Standards in 1970, by Quality of Data and Capital Outlay Coefficients*

| I–O No. | Required Capital Stocks (millions of 1958 $) | | | | Abatement Capital Coefficients (cents per 1958 dollar) |
| | "Good Data" | "Poor Data" | | | |
		Coal-burning	Other	Total	
1	2			2	.005
2			130	130	.463
5		1		1	.050
7			5	5	.130
9		1		1	.048
14		18		18	.020
16		3		3	.017
20			6	6	.056
24	216	32	53	301	1.599
27		45	112	157	.579
28		6		6	.048
29		1		1	.007
31	24		3	27	.191
32		2		2	.013
36	330	28	469	827	7.705
37	818	133	116	1067	3.247
38		77	173	250	1.225
59		4		4	.007
65		2		2	.004
68	1244			1244	2.899
71		3		3	.002
77		3		3	.007
78		18		18	.314
79		21		21	.227
Total	2632	400	1067	4099	

industry gains operating experience with new abatement technologies, cost coefficients for the input-output model may be greatly refined.

REFERENCES

1. Battelle Memorial Institute, "A Cost Analysis of Air Pollution Controls in the Integrated Iron and Steel Industry," prepared for the Division of Economic Effects Research, National Air Pollution Control Administration, U.S. Department of Health, Education and Welfare (May 15, 1969).
2. Clean Air Act (42 USC 1857), part 60, in 36 Fed. Reg. 24876 (Dec. 23, 1971).
3. —— part 420, in 36 Fed. Reg. 15486 (Aug. 14, 1971).
4. Edison Electric Institute, *Statistical Year Book of the Electric Utility Industry for 1969* (New York: 1970).
5. GCA Corporation, "Handbook of Fabric Filter Technology, Vol. I: Fabric Filter Systems Study," prepared for Division of Process Control Engineering, National Air Pollution Control Administration, U.S. Department of Health, Education and Welfare (December 1970).
6. Midwest Research Institute, "Particulate Pollutant System Study, Vol. III: Handbook of Emission Properties," prepared for the Air Pollution Control Office, Environmental Protection Agency (May 1, 1971).
7. Southern Research Institute, "A Manual of Electrostatic Precipitator Technology, Part II: Applications Areas," prepared for Division of Process Control Engineering, National Air Pollution Control Administration, U.S. Department of Health, Education and Welfare (August 25, 1970).
8. U.S. Department of Commerce, Bureau of the Census, *Statistical Abstract of the United States: 1971* (Washington, D.C.: Government Printing Office, 1972).
9. U.S. Department of Commerce, Office of Business Economics, "The Input-Output Structure of the U.S. Economy: 1963," *Survey of Current Business,* 49, No. 11 (November 1969), 16–47.
10. U.S. Department of Health, Education and Welfare, National Air Pollution Control Administration, *Control Techniques for Particulate Air Pollutants* (Washington, D.C.: Government Printing Office, 1970).
11. U.S. Department of the Interior, Bureau of Mines, *Minerals Yearbook, 1968* (Washington, D.C.: Government Printing Office, 1969).
12. U.S. Department of Labor, Bureau of Labor Statistics, "Projections 1970," Bulletin 1536 (Washington, D.C.: Government Printing Office, 1971).
13. —— "Projections of the Post-Vietnam Economy: 1975," Bulletin 1733 (Washington, D.C.: Government Printing Office, 1972).